INTRODUCTION TO MATERIALS SCIENCE

Metals

BONDING

COMMON LATTICE TYPES

GRAIN STRUCTURE AND BOUNDARY

POLYMORPHISM

ALLOYS

IMPERFECTIONS IN METALS

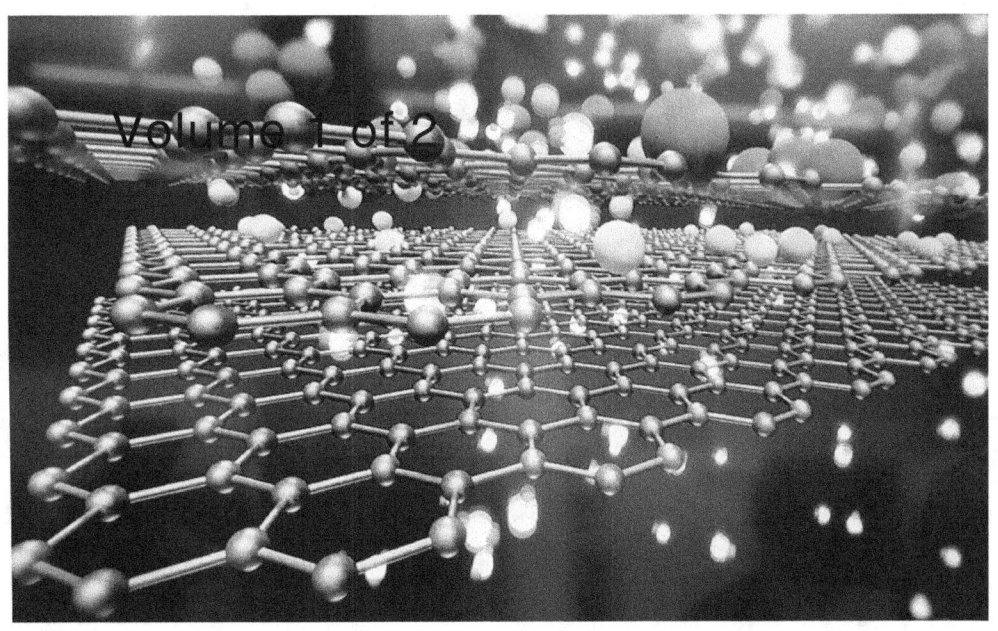

Volume 1 of 2

RED DOT PUBLICATIONS

ABSTRACT

The *Material Science* Handbook was developed to provide a basic understanding of the structure and properties of metals. The handbook includes information on the structure and properties of metals, stress mechanisms in metals, failure modes, and the characteristics of metals. This information will provide you with a foundation for understanding the properties of materials and the way these properties can impose limitations on the operation of equipment and systems.

Key Words: Training Material, Metal Imperfections, Metal Defects, Properties of Metals, Thermal Stress, Thermal Shock, Brittle Fracture, Heat-Up, Cool-Down, Characteristics of Metals

MATERIAL
SCIENCE

OVERVIEW

An understanding of material science will enable personnel to understand why a material was selected for certain applications within their facility.

Almost all processes that take place in the nuclear facilities involve the use of specialized metals. A basic understanding of material science is necessary operators, maintenance personnel, and the technical staff to safely operate and maintain facilities and facility support systems. The information in the handbook is presented to provide a foundation for applying engineering concepts to the job. This knowledge will help personnel more fully understand the impact that their actions may have on the safe and reliable operation of facility components and systems.

The *Material Science* handbook consists of five modules that are contained in two volumes. The following is a brief description of the information presented in each module of the handbook.

Volume 1 of 2

Module 1 - Structure of Metals

Explains the basic structure of metals and how those structures are effected by various processes. The module contains information on the various imperfections and defects that the metal may sustain and how they affect the metal.

Module 2 - Properties of Metals

Contains information on the properties considered when selecting material for a nuclear facility. Each of the properties contains a discussion on how the property is effected and the metal's application.

OVERVIEW (Cont.)

Volume 2 of 2

> Module 3 - Thermal Shock
>
> > Contains material relating to thermal stress and thermal shock effects on a system. Explains how thermal stress and shock combined with pressure can cause major damage to components.
>
> Module 4 - Brittle Fracture
>
> > Contains material on ductile and brittle fracture. These two fractures are the most common in nuclear facilities. Explains how ductile and brittle fracture are effected by the minimum pressurization and temperature curves. Explains the reason why heatup and cooldown rate limits are used when heating up or cooling down the reactor system.
>
> Module 5 - Plant Materials
>
> > Contains information on the commonly used materials and the characteristics desired when selecting material for use.

The information contained in this handbook is by no means all encompassing. An attempt to present the entire subject of material science would be impractical. However, the *Material Science* handbook does present enough information to provide the reader with a fundamental knowledge level sufficient to understand the advanced theoretical concepts presented in other subject areas, and to better understand basic system operation and equipment operations.

MATERIAL SCIENCE
Module 1
Structure of Metals

TABLE OF CONTENTS

LIST OF FIGURES

LIST OF TABLES

REFERENCES

- <u>Academic Program for Nuclear Power Plant Personnel</u>, Volume III, Columbia, MD, General Physics Corporation, Library of Congress Card #A 326517, 1982.

- Foster and Wright, <u>Basic Nuclear Engineering</u>, Fourth Edition, Allyn and Bacon, Inc., 1983.

- Glasstone and Sesonske, <u>Nuclear Reactor Engineering</u>, Third Edition, Van Nostrand Reinhold Company, 1981.

- Metcalfe, Williams, and Castka, <u>Modern Chemistry</u>, Holt, Rinehart, and Winston, New York, NY, 1982.

- <u>Reactor Plant Materials</u>, General Physics Corporation, Columbia Maryland, 1982.

- Savannah River Site, <u>Material Science Course</u>, CS-CRO-IT-FUND-10, Rev. 0, 1991.

- Tweeddale, J.G., <u>The Mechanical Properties of Metals Assessment and Significance</u>, American Elsevier Publishing Company, 1964.

- Weisman, <u>Elements of Nuclear Reactor Design</u>, Elsevier Scientific Publishing Company, 1983.

TERMINAL OBJECTIVE

1.0 Without references, **DESCRIBE** the bonding and patterns that effect the structure of a metal.

ENABLING OBJECTIVES

1.1 **STATE** the five types of bonding that occur in materials and their characteristics.

1.2 **DEFINE** the following terms:

    a. Crystal structure
    b. Body-centered cubic structure
    c. Face-centered cubic structure
    d. Hexagonal close-packed structure

1.3 **STATE** the three lattice-type structures in metals.

1.4 Given a description or drawing, **DISTINGUISH** between the three most common types of crystalline structures.

1.5 **IDENTIFY** the crystalline structure possessed by a metal.

1.6 **DEFINE** the following terms:

    a. Grain
    b. Grain structure
    c. Grain boundary
    d. Creep

1.7 **DEFINE** the term polymorphism.

1.8 **IDENTIFY** the ranges and names for the polymorphism phases associated with uranium metal.

1.9 **IDENTIFY** the polymorphism phase that prevents pure uranium from being used as fuel.

ENABLING OBJECTIVES (Cont.)

1.10   **DEFINE** the term alloy.

1.11   **DESCRIBE** an alloy as to the three possible microstructures and the two general characteristics as compared to pure metals.

1.12   **IDENTIFY** the two desirable properties of type 304 stainless steel.

1.13   **IDENTIFY** the three types of microscopic imperfections found in crystalline structures.

1.14   **STATE** how slip occurs in crystals.

1.15   **IDENTIFY** the four types of bulk defects.

BONDING

The arrangement of atoms in a material determines the behavior and properties of that material. Most of the materials used in the construction of a nuclear reactor facility are metals. In this chapter, we will discuss the various types of bonding that occurs in material selected for use in a reactor facility. The Chemistry Handbook discusses the bonding types in more detail.

EO 1.1 **STATE the five types of bonding that occur in materials and their characteristics.**

Atomic Bonding

Matter, as we know it, exists in three common states. These three states are solid, liquid, and gas. The atomic or molecular interactions that occur within a substance determine its state. In this chapter, we will deal primarily with solids because solids are of the most concern in engineering applications of materials. Liquids and gases will be mentioned for comparative purposes only.

Solid matter is held together by forces originating between neighboring atoms or molecules. These forces arise because of differences in the electron clouds of atoms. In other words, the valence electrons, or those in the outer shell, of atoms determine their attraction for their neighbors. When physical attraction between molecules or atoms of a material is great, the material is held tightly together. Molecules in solids are bound tightly together. When the attractions are weaker, the substance may be in a liquid form and free to flow. Gases exhibit virtually no attractive forces between atoms or molecules, and their particles are free to move independently of each other.

The types of bonds in a material are determined by the manner in which forces hold matter together. Figure 1 illustrates several types of bonds and their characteristics are listed below.

a. Ionic bond - In this type of bond, one or more electrons are wholly transferred from an atom of one element to the atom of the other, and the elements are held together by the force of attraction due to the opposite polarity of the charge.

b. Covalent bond - A bond formed by shared electrons. Electrons are shared when an atom needs electrons to complete its outer shell and can share those electrons with its neighbor. The electrons are then part of both atoms and both shells are filled.

c.   Metallic bond - In this type of bond, the atoms do not share or exchange electrons to bond together. Instead, many electrons (roughly one for each atom) are more or less free to move throughout the metal, so that each electron can interact with many of the fixed atoms.

d.   Molecular bond - When the electrons of neutral atoms spend more time in one region of their orbit, a temporary weak charge will exist. The molecule will weakly attract other molecules. This is sometimes called the van der Waals or molecular bonds.

e.   Hydrogen bond - This bond is similar to the molecular bond and occurs due to the ease with which hydrogen atoms are willing to give up an electron to atoms of oxygen, fluorine, or nitrogen.

Some examples of materials and their bonds are identified in Table 1.

TABLE 1
Examples of Materials and Their Bonds

Material	Bond
Sodium chloride	Ionic
Diamond	Covalent
Sodium	Metallic
Solid H_2	Molecular
Ice	Hydrogen

The type of bond not only determines how well a material is held together, but also determines what microscopic properties the material possesses. Properties such as the ability to conduct heat or electrical current are determined by the freedom of movement of electrons. This is dependent on the type of bonding present. Knowledge of the microscopic structure of a material allows us to predict how that material will behave under certain conditions. Conversely, a material may be synthetically fabricated with a given microscopic structure to yield properties desirable for certain engineering applications.

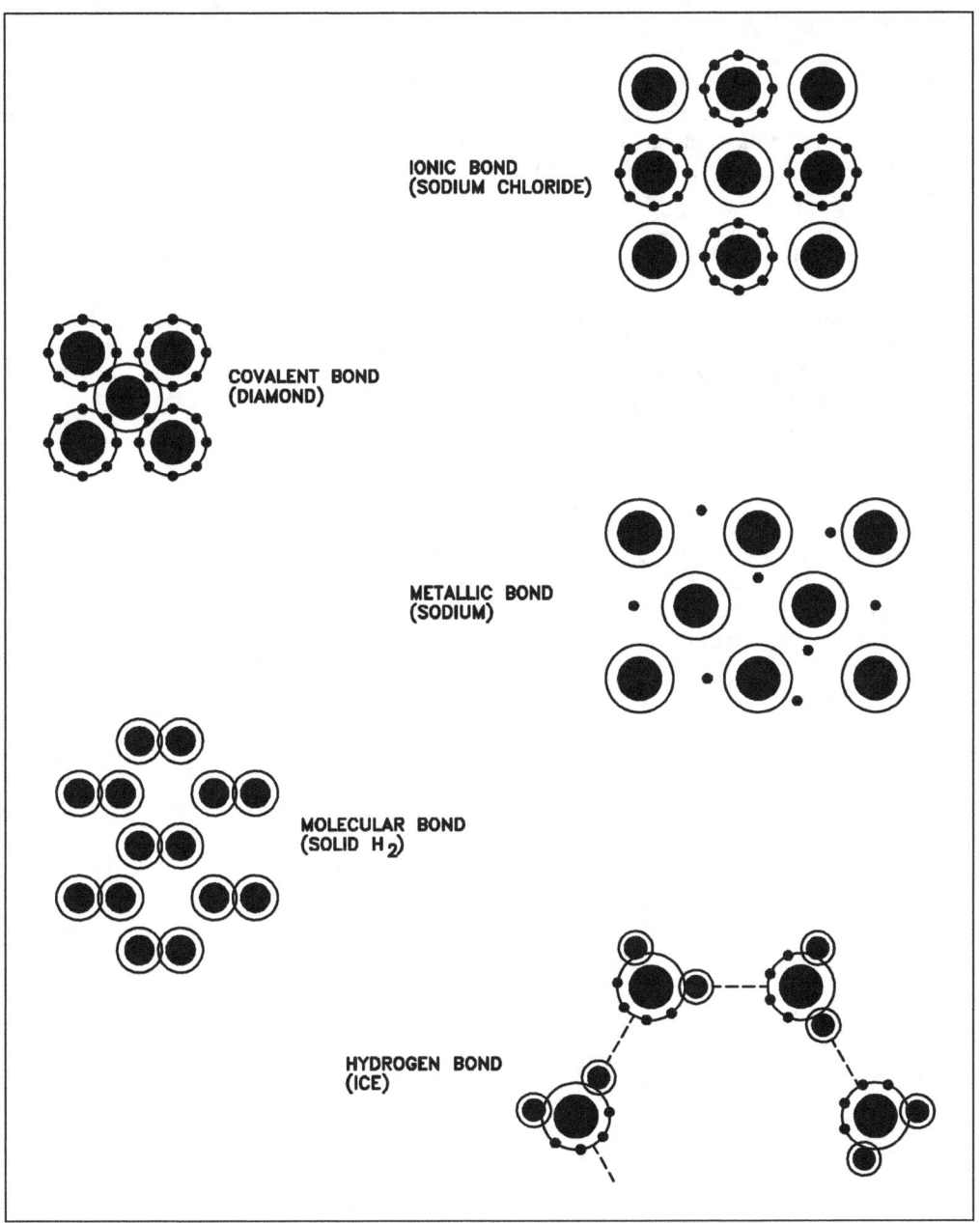

Figure 1 Bonding Types

Order in Microstructures

Solids have greater interatomic attractions than liquids and gases. However, there are wide variations in the properties of solid materials used for engineering purposes. The properties of materials depend on their interatomic bonds. These same bonds also dictate the space between the configuration of atoms in solids. All solids may be classified as either amorphous or crystalline.

Amorphous

Amorphous materials have no regular arrangement of their molecules. Materials like glass and paraffin are considered amorphous. Amorphous materials have the properties of solids. They have definite shape and volume and diffuse slowly. These materials also lack sharply defined melting points. In many respects, they resemble liquids that flow very slowly at room temperature.

Crystalline

In a crystalline structure, the atoms are arranged in a three-dimensional array called a lattice. The lattice has a regular repeating configuration in all directions. A group of particles from one part of a crystal has exactly the same geometric relationship as a group from any other part of the same crystal.

Summary

The important information in this chapter is summarized below.

Bonding Summary

Types of Bonds and Their Characteristics

- Ionic bond - An atom with one or more electrons are wholly transferred from one element to another, and the elements are held together by the force of attraction due to the opposite polarity of the charge.

- Covalent bond - An atom that needs electrons to complete its outer shell shares those electrons with its neighbor.

- Metallic bond - The atoms do not share or exchange electrons to bond together. Instead, many electrons (roughly one for each atom) are more or less free to move throughout the metal, so that each electron can interact with many of the fixed atoms.

- Molecular bond - When neutral atoms undergo shifting in centers of their charge, they can weakly attract other atoms with displaced charges. This is sometimes called the van der Waals bond.

- Hydrogen bond - This bond is similar to the molecular bond and occurs due to the ease with which hydrogen atoms displace their charge.

Order in Microstructures

- Amorphous microstructures lack sharply defined melting points and do not have an orderly arrangement of particles.

- Crystalline microstructures are arranged in three-dimensional arrays called lattices.

COMMON LATTICE TYPES

All metals used in a reactor have crystalline structures. Crystalline microstructures are arranged in three-dimensional arrays called lattices. This chapter will discuss the three most common lattice structures and their characteristics.

EO 1.2   **DEFINE the following terms:**

        a.   **Crystal structure**
        b.   **Body-centered cubic structure**
        c.   **Face-centered cubic structure**
        d.   **Hexagonal close-packed structure**

EO 1.3   **STATE the three lattice-type structures in metals.**

EO 1.4   **Given a description or drawing, DISTINGUISH between the three most common types of crystalline structures.**

EO 1.5   **IDENTIFY the crystalline structure possessed by a metal.**

Common Crystal Structures

In metals, and in many other solids, the atoms are arranged in regular arrays called crystals. A *crystal structure* consists of atoms arranged in a pattern that repeats periodically in a three-dimensional geometric lattice. The forces of chemical bonding causes this repetition. It is this repeated pattern which control properties like strength, ductility, density (described in Module 2, Properties of Metals), conductivity (property of conducting or transmitting heat, electricity, etc.), and shape.

In general, the three most common basic crystal patterns associated with metals are: (a) the body-centered cubic, (b) the face-centered cubic, and (c) the hexagonal close-packed. Figure 2 shows these three patterns.

Body-centered Cubic

In a *body-centered cubic* (BCC) arrangement of atoms, the unit cell consists of eight atoms at the corners of a cube and one atom at the body center of the cube.

Face-centered Cubic

In a *face-centered cubic* (FCC) arrangement of atoms, the unit cell consists of eight atoms at the corners of a cube and one atom at the center of each of the faces of the cube.

Hexagonal Close-packed

In a *hexagonal close-packed* (HCP) arrangement of atoms, the unit cell consists of three layers of atoms. The top and bottom layers contain six atoms at the corners of a hexagon and one atom at the center of each hexagon. The middle layer contains three atoms nestled between the atoms of the top and bottom layers, hence, the name close-packed.

Most diagrams of the structural cells for the BCC and FCC forms of iron are drawn as though they are of the same size, as shown in Figure 2, but they are not. In the BCC arrangement, the structural cell, which uses only nine atoms, is much smaller.

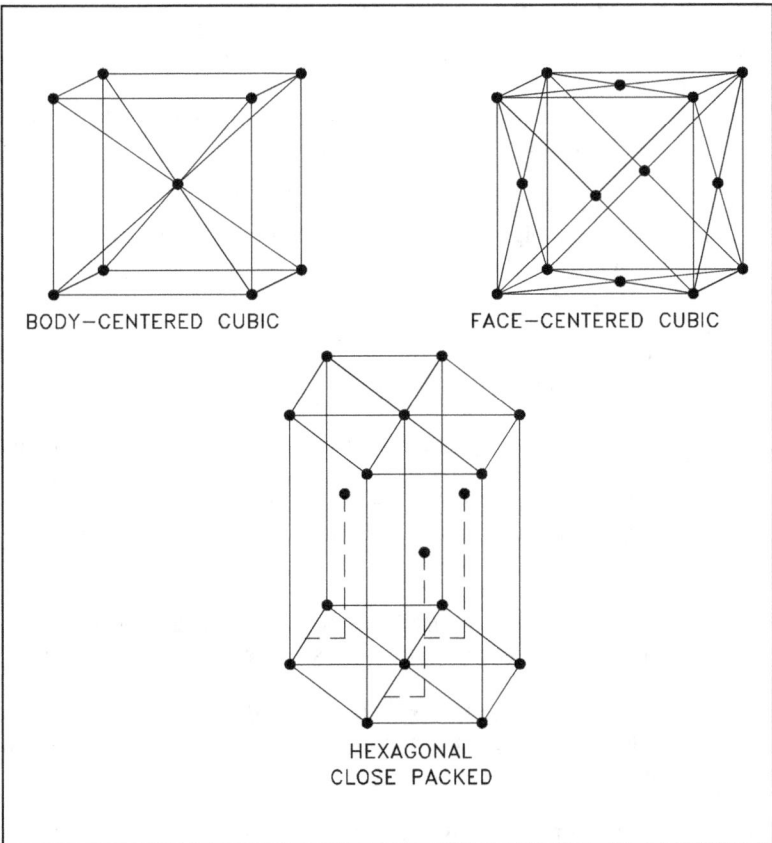

Figure 2 Common Lattice Types

Metals such as α-iron (Fe) (ferrite), chromium (Cr), vanadium (V), molybdenum (Mo), and tungsten (W) possess BCC structures. These BCC metals have two properties in common, high strength and low ductility (which permits permanent deformation). FCC metals such as γ-iron (Fe) (austenite), aluminum (Al), copper (Cu), lead (Pb), silver (Ag), gold (Au), nickel (Ni), platinum (Pt), and thorium (Th) are, in general, of lower strength and higher ductility than BCC metals. HCP structures are found in beryllium (Be), magnesium (Mg), zinc (Zn), cadmium (Cd), cobalt (Co), thallium (Tl), and zirconium (Zr).

Summary

The important information in this chapter is summarized below.

Common Lattice Types Summary

- A crystal structure consists of atoms arranged in a pattern that repeats periodically in a three-dimensional geometric lattice.

- Body-centered cubic structure is an arrangement of atoms in which the unit cell consists of eight atoms at the corners of a cube and one atom at the body center of the cube.

- Face-centered cubic structure is an arrangement of atoms in which the unit cell consists of eight atoms at the corners of a cube and one atom at the center of each of the six faces of the cube.

- Hexagonal close-packed structure is an arrangement of atoms in which the unit cell consists of three layers of atoms. The top and bottom layers contain six atoms at the corners of a hexagon and one atom at the center of each hexagon. The middle layer contains three atoms nestled between the atoms of the top and bottom layers.

- Metals containing BCC structures include ferrite, chromium, vanadium, molybdenum, and tungsten. These metals possess high strength and low ductility.

- Metals containing FCC structures include austenite, aluminum, copper, lead, silver, gold, nickel, platinum, and thorium. These metals possess low strength and high ductility.

- Metals containing HCP structures include beryllium, magnesium, zinc, cadmium, cobalt, thallium, and zirconium. HCP metals are not as ductile as FCC metals.

GRAIN STRUCTURE AND BOUNDARY

Metals contain grains and crystal structures. The individual needs a microscope to see the grains and crystal structures. Grains and grain boundaries help determine the properties of a material.

EO 1.6   **DEFINE the following terms:**

a.   **Grain**
b.   **Grain structure**
c.   **Grain boundary**
d.   **Creep**

Grain Structure and Boundary

If you were to take a small section of a common metal and examine it under a microscope, you would see a structure similar to that shown in Figure 3(a). Each of the light areas is called a *grain*, or crystal, which is the region of space occupied by a continuous crystal lattice. The dark lines surrounding the grains are grain boundaries. The *grain structure* refers to the arrangement of the grains in a metal, with a grain having a particular crystal structure.

The *grain boundary* refers to the outside area of a grain that separates it from the other grains. The grain boundary is a region of misfit between the grains and is usually one to three atom diameters wide. The grain boundaries separate variously-oriented crystal regions (polycrystalline) in which the crystal structures are identical. Figure 3(b) represents four grains of different orientation and the grain boundaries that arise at the interfaces between the grains.

A very important feature of a metal is the average size of the grain. The size of the grain determines the properties of the metal. For example, smaller grain size increases tensile strength and tends to increase ductility. A larger grain size is preferred for improved high-temperature creep properties. *Creep* is the permanent deformation that increases with time under constant load or stress. Creep becomes progressively easier with increasing temperature. Stress and strain are covered in Module 2, Properties of Metals, and creep is covered in Module 5, Plant Materials.

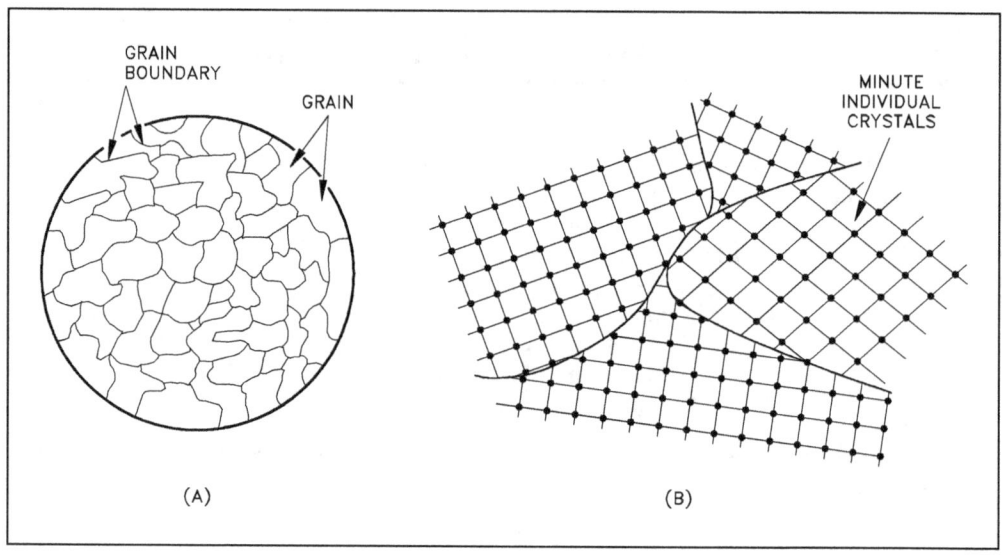

Figure 3 Grains and Boundaries
(a) Microscopic (b) Atomic

Another important property of the grains is their orientation. Figure 4(a) represents a random arrangement of the grains such that no one direction within the grains is aligned with the external boundaries of the metal sample. This random orientation can be obtained by cross rolling the material. If such a sample were rolled sufficiently in one direction, it might develop a grain-oriented structure in the rolling direction as shown in Figure 4(b). This is called preferred orientation. In many cases, preferred orientation is very desirable, but in other instances, it can be most harmful. For example, preferred orientation in uranium fuel elements can result in catastrophic changes in dimensions during use in a nuclear reactor.

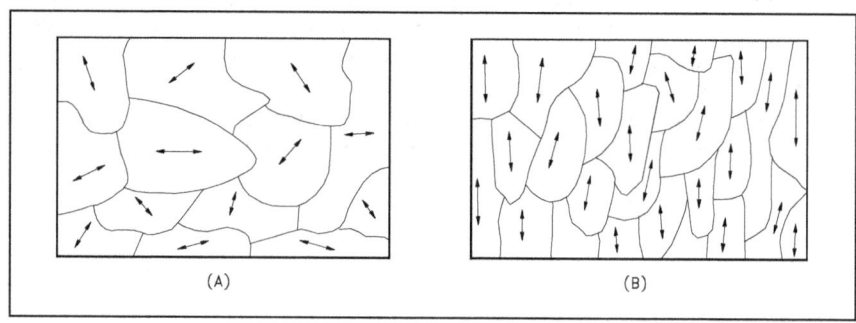

Figure 4 Grain Orientation
(a) Random (b) Preferred

Summary

The important information in this chapter is summarized below.

Grain Structure and Boundary Summary

- Grain is the region of space occupied by a continuous crystal lattice.

- Grain structure is the arrangement of grains in a metal, with a grain having a particular crystal structure.

- Grain boundary is the outside area of grain that separates it from other grains.

- Creep is the permanent deformation that increases with time under constant load or stress.

- Small grain size increases tensile strength and ductility.

POLYMORPHISM

Metals are capable of existing in more than one form at a time. This chapter will discuss this property of metals.

EO 1.7   **DEFINE the term polymorphism.**

EO 1.8   **IDENTIFY the ranges and names for the three polymorphism phases associated with uranium metal.**

EO 1.9   **IDENTIFY the polymorphism phase that prevents pure uranium from being used as fuel.**

Polymorphism Phases

Polymorphism is the property or ability of a metal to exist in two or more crystalline forms depending upon temperature and composition. Most metals and metal alloys exhibit this property. Uranium is a good example of a metal that exhibits polymorphism. Uranium metal can exist in three different crystalline structures. Each structure exists at a specific phase, as illustrated in Figure 5.

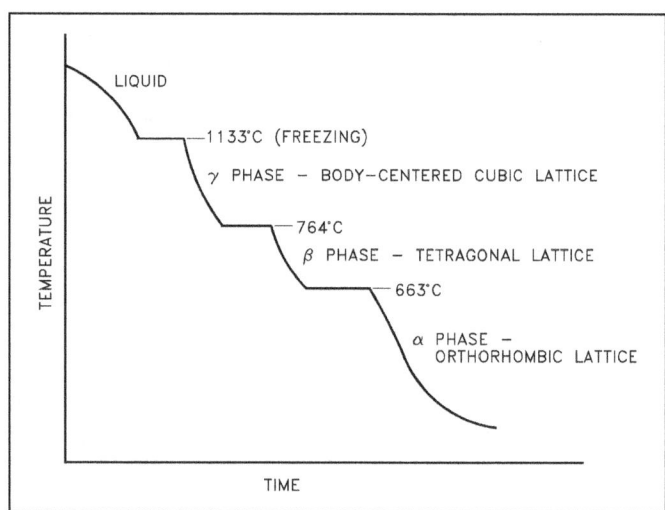

Figure 5 Cooling Curve for Unalloyed Uranium

1.   The alpha phase, from room temperature to 663°C

2.   The beta phase, from 663°C to 764°C

3.   The gamma phase, from 764°C to its melting point of 1133°C

Alpha Phase

The alpha (α) phase is stable at room temperature and has a crystal system characterized by three unequal axes at right angles.

In the alpha phase, the properties of the lattice are different in the X, Y, and Z axes. This is because of the regular recurring state of the atoms is different. Because of this condition, when heated the phase expands in the X and Z directions and shrinks in the Y direction. Figure 6 shows what happens to the dimensions (Å = angstrom, one hundred-millionth of a centimeter) of a unit cell of alpha uranium upon being heated.

As shown, heating and cooling of alpha phase uranium can lead to drastic dimensional changes and gross distortions of the metal. Thus, pure uranium is not used as a fuel, but only in alloys or compounds.

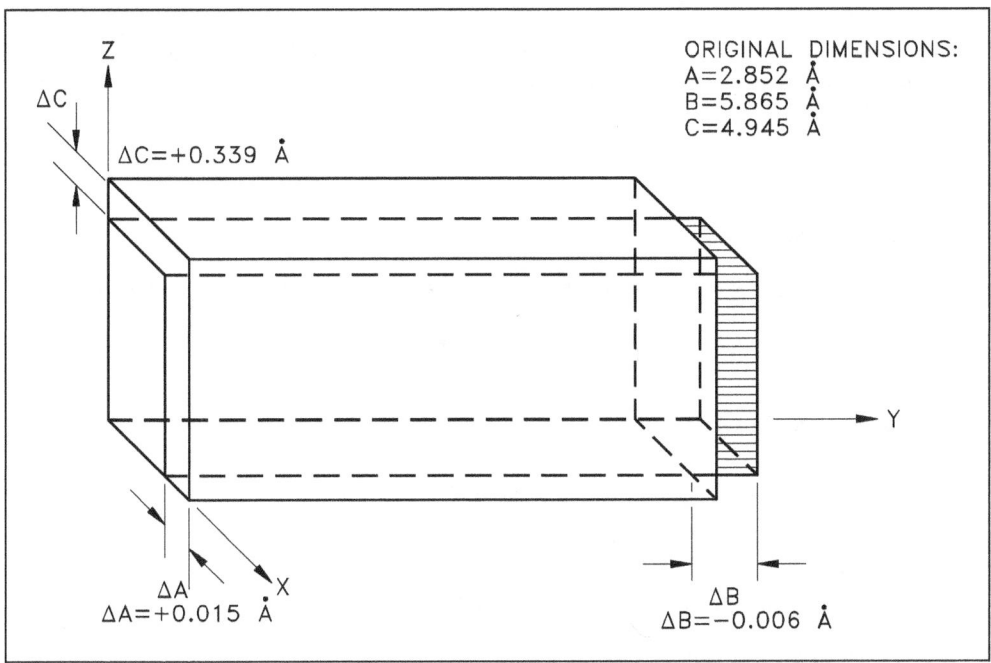

Figure 6 Change in Alpha Uranium Upon Heating From 0 to 300°C

Beta Phase

The beta (β) phase of uranium occurs at elevated temperatures. This phase has a tetragonal (having four angles and four sides) lattice structure and is quite complex.

Gamma Phase

The gamma (γ) phase of uranium is formed at temperatures above those required for beta phase stability. In the gamma phase, the lattice structure is BCC and expands equally in all directions when heated.

Additional Examples

Two additional examples of polymorphism are listed below.

1. Heating iron to 907°C causes a change from BCC (alpha, ferrite) iron to the FCC (gamma, austenite) form.

2. Zirconium is HCP (alpha) up to 863°C, where it transforms to the BCC (beta, zirconium) form.

The properties of one polymorphic form of the same metal will differ from those of another polymorphic form. For example, gamma iron can dissolve up to 1.7% carbon, whereas alpha iron can dissolve only 0.03%.

Summary

The important information in this chapter is summarized below.

Polymorphism Summary

- Polymorphism is the property or ability of a metal to exist in two or more crystalline forms depending upon temperature and composition.

- Metal can exist in three phases or crystalline structures.

- Uranium metal phases are:

    Alpha - Room temperature to 663°C

    Beta - 663°C to 764°C

    Gamma - 764°C to 1133°C

- Alpha phase prevents pure uranium from being used as fuel because of expansion properties.

ALLOYS

Most of the materials used in structural engineering or component fabrication are metals. Alloying is a common practice because metallic bonds allow joining of different types of metals.

EO 1.10   **DEFINE the term alloy.**

EO 1.11   **DESCRIBE an alloy as to the three possible microstructures and the two general characteristics as compared to pure metals.**

EO 1.12   **IDENTIFY the two desirable properties of type 304 stainless steel.**

Alloys

An *alloy* is a mixture of two or more materials, at least one of which is a metal. Alloys can have a microstructure consisting of solid solutions, where secondary atoms are introduced as substitutionals or interstitials (discussed further in the next chapter and Module 5, Plant Materials) in a crystal lattice. An alloy might also be a crystal with a metallic compound at each lattice point. In addition, alloys may be composed of secondary crystals imbedded in a primary polycrystalline matrix. This type of alloy is called a composite (although the term "composite" does not necessarily imply that the component materials are metals). Module 2, Properties of Metals, discusses how different elements change the physical properties of a metal.

Common Characteristics of Alloys

Alloys are usually stronger than pure metals, although they generally offer reduced electrical and thermal conductivity. Strength is the most important criterion by which many structural materials are judged. Therefore, alloys are used for engineering construction. Steel, probably the most common structural metal, is a good example of an alloy. It is an alloy of iron and carbon, with other elements to give it certain desirable properties.

As mentioned in the previous chapter, it is sometimes possible for a material to be composed of several solid phases. The strengths of these materials are enhanced by allowing a solid structure to become a form composed of two interspersed phases. When the material in question is an alloy, it is possible to quench (discussed in more detail in Module 2, Properties of Metals) the metal from a molten state to form the interspersed phases. The type and rate of quenching determines the final solid structure and, therefore, its properties.

Type 304 Stainless Steel

Type 304 stainless steel (containing 18%-20% chromium and 8%-10.5% nickel) is used in the tritium production reactor tanks, process water piping, and original process heat exchangers. This alloy resists most types of corrosion.

Composition of Common Engineering Materials

The wide variety of structures, systems, and components found in DOE nuclear facilities are made from many different types of materials. Many of the materials are alloys with a base metal of iron, nickel, or zirconium. The selection of a material for a specific application is based on many factors including the temperature and pressure that the material will be exposed to, the materials resistance to specific types of corrosion, the materials toughness and hardness, and other material properties.

One material that has wide application in the systems of DOE facilities is stainless steel. There are nearly 40 standard types of stainless steel and many other specialized types under various trade names. Through the modification of the kinds and quantities of alloying elements, the steel can be adapted to specific applications. Stainless steels are classified as austenitic or ferritic based on their lattice structure. Austenitic stainless steels, including 304 and 316, have a face-centered cubic structure of iron atoms with the carbon in interstitial solid solution. Ferritic stainless steels, including type 405, have a body-centered cubic iron lattice and contain no nickel. Ferritic steels are easier to weld and fabricate and are less susceptible to stress corrosion cracking than austenitic stainless steels. They have only moderate resistance to other types of chemical attack.

Other metals that have specific applications in some DOE nuclear facilities are inconel and zircaloy. The composition of these metals and various types of stainless steel are listed in Table 2 below.

	%Fe	%C Max	%Cr	%Ni	%Mo	%Mn Max	%Si Max	%Zr
TABLE 2								
Typical Composition of Common Engineering Materials								
304 Stainless Steel	Bal.	0.08	19	10		2	1	
304L Stainless Steel	Bal.	0.03	18	8		2	1	
316 Stainless Steel	Bal.	0.08	17	12	2.5	2	1	
316L Stainless Steel	Bal.	0.03	17	12	2.5	2		
405 Stainless Steel	Bal.	0.08	13			1	1	
Inconel	8	0.15	15	Bal.		1	0.5	
Zircaloy-4	0.21		0.1					Bal.

Summary

The important information in this chapter is summarized below.

Alloys Summary

- An alloy is a mixture of two or more materials, at least one of which is a metal.

- Alloy microstructures

  Solid solutions, where secondary atoms introduced as substitutionals or interstitials in a crystal lattice.

  Crystal with metallic bonds

  Composites, where secondary crystals are imbedded in a primary polycrystalline matrix.

- Alloys are usually stronger than pure metals although alloys generally have reduced electrical and thermal conductivities than pure metals.

- The two desirable properties of type 304 stainless steel are corrosion resistance and high toughness.

IMPERFECTIONS IN METALS

The discussion of order in microstructures in the previous chapters assumed idealized microstructures. In reality, materials are not composed of perfect crystals, nor are they free of impurities that alter their properties. Even amorphous solids have imperfections and impurities that change their structure.

EO 1.13   **IDENTIFY the three types of microscopic imperfections found in crystalline structures.**

EO 1.14   **STATE how slip occurs in crystals.**

EO 1.15   **IDENTIFY the four types of bulk defects.**

Microscopic Imperfections

Microscopic imperfections are generally classified as either point, line, or interfacial imperfections.

1.   Point imperfections have atomic dimensions.

2.   Line imperfections or dislocations are generally many atoms in length.

3.   Interfacial imperfections are larger than line defects and occur over a two-dimensional area.

Point Imperfections

Point imperfections in crystals can be divided into three main defect categories. They are illustrated in Figure 7.

1.   Vacancy defects result from a missing atom in a lattice position. The vacancy type of defect can result from imperfect packing during the crystallization process, or it may be due to increased thermal vibrations of the atoms brought about by elevated temperature.

2.   Substitutional defects result from an impurity present at a lattice position.

3.   Interstitial defects result from an impurity located at an interstitial site or one of the lattice atoms being in an interstitial position instead of being at its lattice position. Interstitial refers to locations between atoms in a lattice structure.

Interstitial impurities called network modifiers act as point defects in amorphous solids. The presence of point defects can enhance or lessen the value of a material for engineering construction depending upon the intended use.

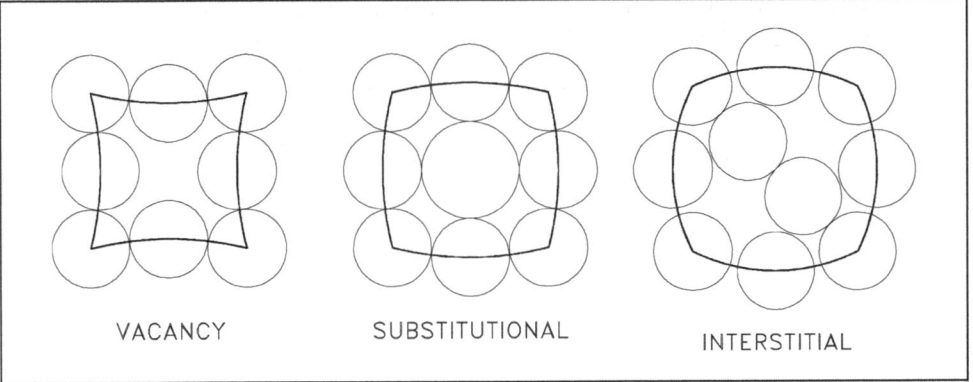

Figure 7 Point Defects

Line Imperfections

Line imperfections are called dislocations and occur in crystalline materials only. Dislocations can be an edge type, screw type, or mixed type, depending on how they distort the lattice, as shown in Figure 8. It is important to note that dislocations cannot end inside a crystal. They must end at a crystal edge or other dislocation, or they must close back on themselves.

Edge dislocations consist of an extra row or plane of atoms in the crystal structure. The imperfection may extend in a straight line all the way through the crystal or it may follow an irregular path. It may also be short, extending only a small distance into the crystal causing a slip of one atomic distance along the glide plane (direction the edge imperfection is moving).

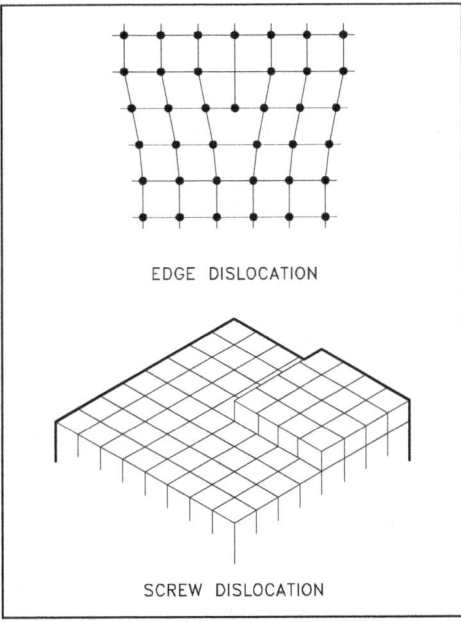

Figure 8 Line Defects (Dislocations)

The slip occurs when the crystal is subjected to a stress, and the dislocation moves through the crystal until it reaches the edge or is arrested by another dislocation, as shown in Figure 9. Position 1 shows a normal crystal structure. Position 2 shows a force applied from the left side and a counterforce applied from the right side. Positions 3 to 5 show how the structure is slipping. Position 6 shows the final deformed crystal structure. The slip of one active plane is ordinarily on the order of 1000 atomic distances and, to produce yielding, slip on many planes is required.

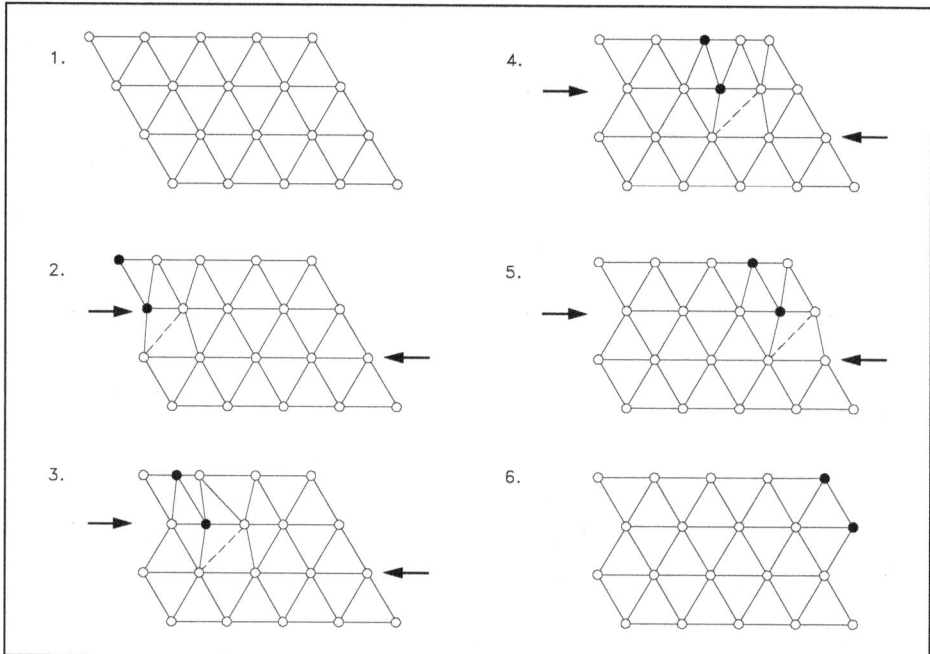

Figure 9 Slips

Screw dislocations can be produced by a tearing of the crystal parallel to the slip direction. If a screw dislocation is followed all the way around a complete circuit, it would show a slip pattern similar to that of a screw thread. The pattern may be either left or right handed. This requires that some of the atomic bonds are re-formed continuously so that the crystal has almost the same form after yielding that it had before.

The orientation of dislocations may vary from pure edge to pure screw. At some intermediate point, they may possess both edge and screw characteristics. The importance of dislocations is based on the ease at which they can move through crystals.

Interfacial Imperfections

Interfacial imperfections exist at an angle between any two faces of a crystal or crystal form. These imperfections are found at free surfaces, domain boundaries, grain boundaries, or interphase boundaries. Free surfaces are interfaces between gases and solids. Domain boundaries refer to interfaces where electronic structures are different on either side causing each side to act differently although the same atomic arrangement exists on both sides. Grain boundaries exist between crystals of similar lattice structure that possess different spacial orientations. Polycrystalline materials are made up of many grains which are separated by distances typically of several atomic diameters. Finally, interphase boundaries exist between the regions where materials exist in different phases (i.e., BCC next to FCC structures).

Macroscopic Defects

Three-dimensional macroscopic defects are called bulk defects. They generally occur on a much larger scale than the microscopic defects. These macroscopic defects generally are introduced into a material during refinement from its raw state or during fabrication processes.

The most common bulk defect arises from foreign particles being included in the prime material. These second-phase particles, called inclusions, are seldom wanted because they significantly alter the structural properties. An example of an inclusion may be oxide particles in a pure metal or a bit of clay in a glass structure.

Other bulk defects include gas pockets or shrinking cavities found generally in castings. These spaces weaken the material and are therefore guarded against during fabrication. The working and forging of metals can cause cracks that act as stress concentrators and weaken the material. Any welding or joining defects may also be classified as bulk defects.

Summary

The important information in this chapter is summarized below.

Imperfections in Metals Summary

Microscopic Imperfections

•   Point imperfections are in the size range of individual atoms.

•   Line (dislocation) imperfections are generally many atoms in length. Line imperfections can be of the edge type, screw type, or mixed type, depending on lattice distortion. Line imperfections cannot end inside a crystal; they must end at crystal edge or other dislocation, or close back on themselves.

•   Interfacial imperfections are larger than line imperfections and occur over a two dimensional area. Interfacial imperfections exist at free surfaces, domain boundaries, grain boundaries, or interphase boundaries.

•   Slip occurs when a crystal is subjected to stress and the dislocations march through the crystal until they reach the edge or are arrested by another dislocation.

Macroscopic Defects

•   Bulk defects are three dimensional defects.

        Foreign particles included in the prime material (inclusions) are most common bulk defect

        Gas pockets

        Shrinking cavities

        Welding or joining defects

Department of Energy
Fundamentals Handbook

MATERIAL SCIENCE
Module 2
Properties of Metals

TABLE OF CONTENTS

TABLE OF CONTENTS (Cont.)

TABLE OF CONTENTS (Cont.)

LIST OF FIGURES

LIST OF TABLES

REFERENCES

•   <u>Academic Program for Nuclear Power Plant Personnel</u>, Volume III, Columbia, MD, General Physics Corporation, Library of Congress Card #A 326517, 1982.

•   Berry, <u>Corrosion Problems in Light Water Nuclear Reactors 1984</u>, Speller Award Lecture, presented during CORROSION/84, April 1984, New Orleans, Louisiana.

•   Foster and Wright, <u>Basic Nuclear Engineering</u>, Fourth Edition, Allyn and Bacon, Inc., 1983.

•   Glasstone and Sesonske, <u>Nuclear Reactor Engineering</u>, Third Edition, Van Nostrand Reinhold Company, 1981.

•   Makansi, <u>Solving Power Plant Corrosion Problems</u>, Power Special Report, 1983.

•   McKay, <u>Mechanisms of Denting in Nuclear Steam Generators</u>, presented during CORROSION/82, Paper 214, March 1982, Houston, Texas.

•   Owens, <u>Stress Corrosion Cracking</u>, presented during CORROSION/85, Paper No. 93, NACE, Houston, Texas, 1985.

•   Raymond, <u>Hydrogen Embrittlement Control</u>, ASTM, Standardization News, December 1985.

•   <u>Reactor Plant Materials</u>, General Physics Corporation, Columbia Maryland, 1982.

•   Savannah River Site, <u>Material Science Course</u>, CS-CRO-IT-FUND-10, Rev. 0, 1991.

•   Tweeddale, J.G., <u>The Mechanical Properties of Metals Assessment and Significance</u>, American Elsevier Publishing Company, 1964.

•   Weisman, <u>Elements of Nuclear Reactor Design</u>, Elsevier Scientific Publishing Company, 1983.

TERMINAL OBJECTIVE

1.0   Without references, **DESCRIBE** how changes in stress, strain, and physical and chemical properties effect the materials used in a reactor plant.

ENABLING OBJECTIVES

1.1   **DEFINE** the following terms:

    a.   Stress
    b.   Tensile stress
    c.   Compressive stress
    d.   Shear stress
    e.   Compressibility

1.2   **DISTINGUISH** between the following types of stresses by the direction in which stress is applied.

    a.   Tensile
    b.   Compressive
    c.   Shear

1.3   **DEFINE** the following terms:

    a.   Strain
    b.   Plastic deformation
    c.   Proportional limit

1.4   **IDENTIFY** the two common forms of strain.

1.5   **DISTINGUISH** between the two common forms of strain as to dimensional change.

1.6   **STATE** how iron crystalline lattice, γ and α, structure deforms under load.

1.7   **STATE** Hooke's Law.

1.8   **DEFINE** Young's Modulus (Elastic Modulus) as it relates to stress.

ENABLING OBJECTIVES (Cont.)

1.9   Given the values of the associated material properties, **CALCULATE** the elongation of a material using Hooke's Law.

1.10   **DEFINE** the following terms:

      a.   Bulk Modulus
      b.   Fracture point

1.11   Given stress-strain curves for ductile and brittle material, **IDENTIFY** the following specific points on a stress-strain curve.

      a.   Proportional limit
      b.   Yield point
      c.   Ultimate strength
      d.   Fracture point

1.12   Given a stress-strain curve, **IDENTIFY** whether the type of material represented is ductile or brittle.

1.13   Given a stress-strain curve, **INTERPRET** a stress-strain curve for the following:

      a.   Application of Hooke's Law
      b.   Elastic region
      c.   Plastic region

1.14   **DEFINE** the following terms:

      a.   Strength
      b.   Ultimate tensile strength
      c.   Yield strength
      d.   Ductility
      e.   Malleability
      f.   Toughness
      g.   Hardness

1.15   **IDENTIFY** how slip effects the strength of a metal.

ENABLING OBJECTIVES (Cont.)

1.16   **DESCRIBE** the effects on ductility caused by:

   a.   Temperature changes
   b.   Irradiation
   c.   Cold working

1.17   **IDENTIFY** the reactor plant application for which high ductility is desirable.

1.18   **STATE** how heat treatment effects the properties of heat-treated steel and carbon steel.

1.19   **DESCRIBE** the adverse effects of welding on metal including types of stress and method(s) for minimizing stress.

1.20   **STATE** the reason that galvanic corrosion is a concern in design and material selection.

1.21   **DESCRIBE** hydrogen embrittlement including the two required conditions and the formation process.

1.22   **IDENTIFY** why zircaloy-4 is less susceptible to hydrogen embrittlement than zircaloy-2.

Intentionally Left Blank

STRESS

Any component, no matter how simple or complex, has to transmit or sustain a mechanical load of some sort. The load may be one of the following types: a load that is applied steadily ("dead" load); a load that fluctuates, with slow or fast changes in magnitude ("live" load); a load that is applied suddenly (shock load); or a load due to impact in some form. Stress is a form of load that may be applied to a component. Personnel need to be aware how stress may be applied and how it effects the component.

EO 1.1   **DEFINE the following terms:**

a.   **Stress**
b.   **Tensile stress**
c.   **Compressive stress**
d.   **Shear stress**
e.   **Compressibility**

EO 1.2   **DISTINGUISH between the following types of stresses by the direction in which stress is applied.**

a.   **Tensile**
b.   **Compressive**
c.   **Shear**

Definition of Stress

When a metal is subjected to a load (force), it is distorted or deformed, no matter how strong the metal or light the load. If the load is small, the distortion will probably disappear when the load is removed. The intensity, or degree, of distortion is known as *strain*. If the distortion disappears and the metal returns to its original dimensions upon removal of the load, the strain is called *elastic strain*. If the distortion disappears and the metal remains distorted, the strain type is called *plastic strain*. Strain will be discussed in more detail in the next chapter.

When a load is applied to metal, the atomic structure itself is strained, being compressed, warped or extended in the process. The atoms comprising a metal are arranged in a certain geometric pattern, specific for that particular metal or alloy, and are maintained in that pattern by interatomic forces. When so arranged, the atoms are in their state of minimum energy and tend to remain in that arrangement. Work must be done on the metal (that is, energy must be added) to distort the atomic pattern. (Work is equal to force times the distance the force moves.)

Stress is the internal resistance, or counterfource, of a material to the distorting effects of an external force or load. These counterforces tend to return the atoms to their normal positions. The total resistance developed is equal to the external load. This resistance is known as *stress*.

Although it is impossible to measure the intensity of this stress, the external load and the area to which it is applied can be measured. Stress (σ) can be equated to the load per unit area or the force (F) applied per cross-sectional area (A) perpendicular to the force as shown in Equation (2-1).

$$\text{Stress} = \sigma = \frac{F}{A} \tag{2-1}$$

where:

σ = stress (psi or lbs of force per in.^{2})

F = applied force (lbs of force per in.^{2})

A = cross-sectional area (in.^{2})

Types of Stress

Stresses occur in any material that is subject to a load or any applied force. There are many types of stresses, but they can all be generally classified in one of six categories: residual stresses, structural stresses, pressure stresses, flow stresses, thermal stresses, and fatigue stresses.

Residual Stresses

Residual stresses are due to the manufacturing processes that leave stresses in a material. Welding leaves residual stresses in the metals welded. Stresses associated with welding are further discussed later in this module.

Structural Stresses

Structural stresses are stresses produced in structural members because of the weights they support. The weights provide the loadings. These stresses are found in building foundations and frameworks, as well as in machinery parts.

Pressure Stresses

Pressure stresses are stresses induced in vessels containing pressurized materials. The loading is provided by the same force producing the pressure. In a reactor facility, the reactor vessel is a prime example of a pressure vessel.

Flow Stresses

Flow stresses occur when a mass of flowing fluid induces a dynamic pressure on a conduit wall. The force of the fluid striking the wall acts as the load. This type of stress may be applied in an unsteady fashion when flow rates fluctuate. Water hammer is an example of a transient flow stress.

Thermal Stresses

Thermal stresses exist whenever temperature gradients are present in a material. Different temperatures produce different expansions and subject materials to internal stress. This type of stress is particularly noticeable in mechanisms operating at high temperatures that are cooled by a cold fluid. Thermal stress is further discussed in Module 3.

Fatigue Stresses

Fatigue stresses are due to cyclic application of a stress. The stresses could be due to vibration or thermal cycling. Fatigue stresses are further discussed in Module 4.

The importance of all stresses is increased when the materials supporting them are flawed. Flaws tend to add additional stress to a material. Also, when loadings are cyclic or unsteady, stresses can effect a material more severely. The additional stresses associated with flaws and cyclic loading may exceed the stress necessary for a material to fail.

Types of Applied Stress

Stress intensity within the body of a component is expressed as one of three basic types of internal load. They are known as tensile, compressive, and shear. Figure 1 illustrates the different types of stress. Mathematically, there are only two types of internal load because tensile and compressive stress may be regarded as the positive and negative versions of the same type of normal loading.

However, in mechanical design, the response of components to the two conditions can be so different that it is better, and safer, to regard them as separate types.

As illustrated in Figure 1, the plane of a tensile or compressive stress lies perpendicular to the axis of operation of the force from which it originates. The plane of a shear stress lies in the plane of the force system from which it originates. It is essential to keep these differences quite clear both in mind and mode of expression.

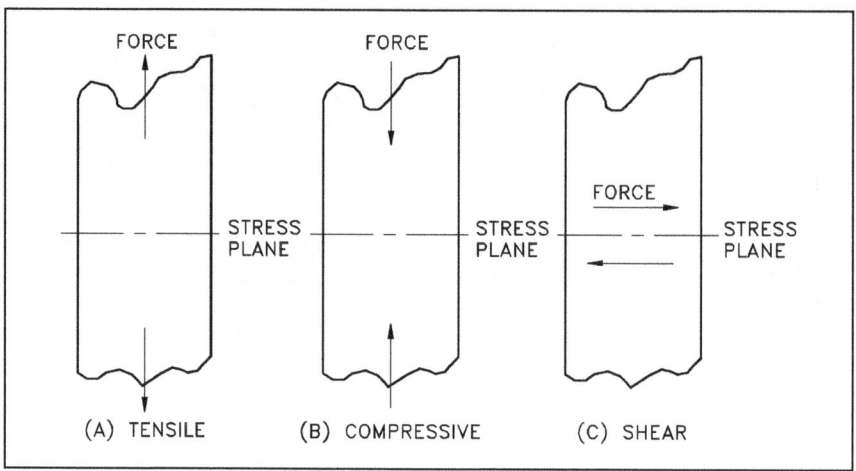

Figure 1 Types of Applied Stress

Tensile Stress

Tensile stress is that type of stress in which the two sections of material on either side of a stress plane tend to pull apart or elongate as illustrated in Figure 1(a).

Compressive Stress

Compressive stress is the reverse of tensile stress. Adjacent parts of the material tend to press against each other through a typical stress plane as illustrated in Figure 1(b).

Shear Stress

Shear stress exists when two parts of a material tend to slide across each other in any typical plane of shear upon application of force parallel to that plane as illustrated in Figure 1(c).

Assessment of mechanical properties is made by addressing the three basic stress types. Because tensile and compressive loads produce stresses that act across a plane, in a direction perpendicular (normal) to the plane, tensile and compressive stresses are called normal stresses. The shorthand designations are as follows.

For tensile stresses: "$+S_N$" (or "$S_{N''}$") or "σ" (sigma)

For compressive stresses: "$-S_N$" or "$-\sigma$" (minus sigma)

The ability of a material to react to compressive stress or pressure is called *compressibility*. For example, metals and liquids are incompressible, but gases and vapors are compressible. The shear stress is equal to the force divided by the area of the face parallel to the direction in which the force acts, as shown in Figure 1(c).

Two types of stress can be present simultaneously in one plane, provided that one of the stresses is shear stress. Under certain conditions, different basic stress type combinations may be simultaneously present in the material. An example would be a reactor vessel during operation. The wall has tensile stress at various locations due to the temperature and pressure of the fluid acting on the wall. Compressive stress is applied from the outside at other locations on the wall due to outside pressure, temperature, and constriction of the supports associated with the vessel. In this situation, the tensile and compressive stresses are considered principal stresses. If present, shear stress will act at a 90° angle to the principal stress.

Summary

The important information in this chapter is summarized below.

Stress Summary

- Stress is the internal resistance of a material to the distorting effects of an external force or load.

$$\text{Stress} = \sigma = \frac{F}{A}$$

- Three types of stress

    Tensile stress is the type of stress in which the two sections of material on either side of a stress plane tend to pull apart or elongate.

    Compressive stress is the reverse of tensile stress. Adjacent parts of the material tend to press against each other.

    Shear stress exists when two parts of a material tend to slide across each other upon application of force parallel to that plane.

- Compressibility is the ability of a material to react to compressive stress or pressure.

STRAIN

When stress is present strain will be involved also. The two types of strain will be discussed in this chapter. Personnel need to be aware how strain may be applied and how it affects the component.

EO 1.3   **DEFINE the following terms:**

               a.   **Strain**
               b.   **Plastic deformation**
               c.   **Proportional limit**

EO 1.4   **IDENTIFY the two common forms of strain.**

EO 1.5   **DISTINGUISH between the two common forms of strain according to dimensional change.**

EO 1.6   **STATE how iron crystalline lattice structures, γ and α, deform under load.**

Definition of Strain

In the use of metal for mechanical engineering purposes, a given state of stress usually exists in a considerable volume of the material. Reaction of the atomic structure will manifest itself on a macroscopic scale. Therefore, whenever a stress (no matter how small) is applied to a metal, a proportional dimensional change or distortion must take place.

Such a proportional dimensional change (intensity or degree of the distortion) is called *strain* and is measured as the total elongation per unit length of material due to some applied stress. Equation 2-2 illustrates this proportion or distortion.

$$\text{Strain} = \varepsilon = \frac{\delta}{L} \tag{2-2}$$

where:

ε = strain (in./in.)
δ = total elongation (in.)
L = original length (in.)

Types of Strain

Strain may take two forms; elastic strain and plastic deformation.

Elastic Strain

Elastic strain is a transitory dimensional change that exists only while the initiating stress is applied and disappears immediately upon removal of the stress. Elastic strain is also called elastic deformation. The applied stresses cause the atoms in a crystal to move from their equilibrium position. All the atoms are displaced the same amount and still maintain their relative geometry. When the stresses are removed, all the atoms return to their original positions and no permanent deformation occurs.

Plastic Deformation

Plastic deformation (or plastic strain) is a dimensional change that does not disappear when the initiating stress is removed. It is usually accompanied by some elastic strain.

The phenomenon of elastic strain and plastic deformation in a material are called *elasticity* and *plasticity*, respectively.

At room temperature, most metals have some elasticity, which manifests itself as soon as the slightest stress is applied. Usually, they also possess some plasticity, but this may not become apparent until the stress has been raised appreciably. The magnitude of plastic strain, when it does appear, is likely to be much greater than that of the elastic strain for a given stress increment. Metals are likely to exhibit less elasticity and more plasticity at elevated temperatures. A few pure unalloyed metals (notably aluminum, copper and gold) show little, if any, elasticity when stressed in the annealed (heated and then cooled slowly to prevent brittleness) condition at room temperature, but do exhibit marked plasticity. Some unalloyed metals and many alloys have marked elasticity at room temperature, but no plasticity.

The state of stress just before plastic strain begins to appear is known as the *proportional limit*, or elastic limit, and is defined by the stress level and the corresponding value of elastic strain. The proportional limit is expressed in pounds per square inch. For load intensities beyond the proportional limit, the deformation consists of both elastic and plastic strains.

As mentioned previously in this chapter, strain measures the proportional dimensional change with no load applied. Such values of strain are easily determined and only cease to be sufficiently accurate when plastic strain becomes dominant.

When metal experiences strain, its volume remains constant. Therefore, if volume remains constant as the dimension changes on one axis, then the dimensions of at least one other axis must change also. If one dimension increases, another must decrease. There are a few exceptions. For example, *strain hardening* involves the absorption of strain energy in the material structure, which results in an increase in one dimension without an offsetting decrease in other dimensions. This causes the density of the material to decrease and the volume to increase.

If a tensile load is applied to a material, the material will elongate on the axis of the load (perpendicular to the tensile stress plane), as illustrated in Figure 2(a). Conversely, if the load is compressive, the axial dimension will decrease, as illustrated in Figure 2(b). If volume is constant, a corresponding lateral contraction or expansion must occur. This lateral change will bear a fixed relationship to the axial strain. The relationship, or ratio, of lateral to axial strain is called *Poisson's ratio* after the name of its discoverer. It is usually symbolized by v.

Deformation of Cubic Structures

Whether or not a material can deform plastically at low applied stresses depends on its lattice structure. It is easier for planes of atoms to slide by each other if those planes are closely packed. Therefore lattice structures with closely packed planes allow more plastic deformation than those that are not closely packed. Also, cubic lattice structures allow slippage to occur more easily than non-cubic lattices. This is because of their symmetry which provides closely packed planes in several directions. Most metals are made of the body-centered cubic (BCC), face-centered cubic (FCC), or hexagonal close-packed (HCP) crystals, discussed in more detail in the Module 1, Structure of Metals. A face-centered cubic crystal structure will deform more readily under load before breaking than a body-centered cubic structure.

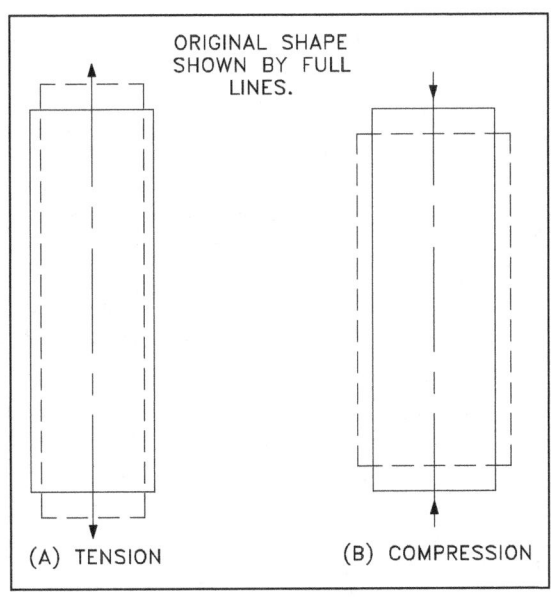

Figure 2 Change of Shape of Cylinder Under Stress

The BCC lattice, although cubic, is not closely packed and forms strong metals. α-iron and tungsten have the BCC form. The FCC lattice is both cubic and closely packed and forms more ductile materials. γ-iron, silver, gold, and lead are FCC structured. Finally, HCP lattices are closely packed, but not cubic. HCP metals like cobalt and zinc are not as ductile as the FCC metals.

Summary

The important information in this chapter is summarized below.

Strain Summary

- Strain is the proportional dimensional change, or the intensity or degree of distortion, in a material under stress.

- Plastic deformation is the dimensional change that does not disappear when the initiating stress is removed.

- Proportional limit is the amount of stress just before the point (threshold) at which plastic strain begins to appear or the stress level and the corresponding value of elastic strain.

- Two types of strain:

  > Elastic strain is a transitory dimensional change that exists only while the initiating stress is applied and disappears immediately upon removal of the stress.

  > Plastic strain (plastic deformation) is a dimensional change that does not disappear when the initiating stress is removed.

- γ-iron face-centered cubic crystal structures deform more readily under load before breaking than α-iron body-centered cubic structures.

YOUNG'S MODULUS

This chapter discusses the mathematical method used to calculate the elongation of a material under tensile force and elasticity of a material.

EO 1.7   **STATE Hooke's Law.**

EO 1.8   **DEFINE Young's Modulus (Elastic Modulus) as it relates to stress.**

EO 1.9   **Given the values of the associated material properties, CALCULATE the elongation of a material using Hooke's Law.**

Hooke's Law

If a metal is lightly stressed, a temporary deformation, presumably permitted by an elastic displacement of the atoms in the space lattice, takes place. Removal of the stress results in a gradual return of the metal to its original shape and dimensions. In 1678 an English scientist named Robert Hooke ran experiments that provided data that showed that in the elastic range of a material, strain is proportional to stress. The elongation of the bar is directly proportional to the tensile force and the length of the bar and inversely proportional to the cross-sectional area and the modulus of elasticity.

Hooke's experimental law may be given by Equation (2-3).

$$\delta = \frac{P\ell}{AE} \qquad (2-3)$$

This simple linear relationship between the force (stress) and the elongation (strain) was formulated using the following notation.

P   =   force producing extension of bar (lbf)
ℓ =   length of bar (in.)
A   =   cross-sectional area of bar (in.^{2})
δ =   total elongation of bar (in.)
E   =   elastic constant of the material, called the Modulus of Elasticity, or Young's Modulus (lbf/in.^{2})

The quantity E, the ratio of the unit stress to the unit strain, is the modulus of elasticity of the material in tension or compression and is often called *Young's Modulus.*

Previously, we learned that tensile stress, or simply stress, was equated to the load per unit area or force applied per cross-sectional area perpendicular to the force measured in pounds force per square inch.

$$\sigma = \frac{P}{A} \qquad (2\text{-}4)$$

We also learned that tensile strain, or the elongation of a bar per unit length, is determined by:

$$\varepsilon = \frac{\delta}{\ell} \qquad (2\text{-}5)$$

Thus, the conditions of the experiment described above are adequately expressed by Hooke's Law for elastic materials. For materials under tension, strain (ε) is proportional to applied stress σ.

$$\varepsilon = \frac{\sigma}{E} \qquad (2\text{-}6)$$

where

$$E \quad = \quad \text{Young's Modulus (lbf/in.}^2)$$

$$\sigma \quad = \quad \text{stress (psi)}$$

$$\varepsilon \quad = \quad \text{strain (in./in.)}$$

Young's Modulus (Elastic Modulus)

Young's Modulus (sometimes referred to as Modulus of Elasticity, meaning "measure" of elasticity) is an extremely important characteristic of a material. It is the numerical evaluation of Hooke's Law, namely the ratio of stress to strain (the measure of resistance to elastic deformation). To calculate Young's Modulus, stress (at any point) below the proportional limit is divided by corresponding strain. It can also be calculated as the slope of the straight-line portion of the stress-strain curve. (The positioning on a stress-strain curve will be discussed later.)

$$E = \text{Elastic Modulus} = \frac{\text{stress}}{\text{strain}} = \frac{\text{psi}}{\text{in./in.}} = \text{psi}$$

or

$$E = \frac{\sigma}{\varepsilon} \qquad (2\text{-}7)$$

We can now see that Young's Modulus may be easily calculated, provided that the stress and corresponding unit elongation or strain have been determined by a tensile test as described previously. Strain (ε) is a number representing a ratio of two lengths; therefore, we can conclude that the Young's Modulus is measured in the same units as stress (σ), that is, in pounds per square inch. Table 1 gives average values of the Modulus E for several metals used in DOE facilities construction. Yield strength and ultimate strength will be discussed in more detail in the next chapter.

TABLE 1
Properties of Common Structural Materials

	E (psi)	Yield Strength (psi)	Ultimate Strength (psi)
Aluminum	1.0×10^7	3.5×10^4 to 4.5×10^4	5.4×10^4 to 6.5×10^4
Stainless Steel	2.9×10^7	4.0×10^4 to 5.0×10^4	7.8×10^4 to 10×10^4
Carbon Steel	3.0×10^7	3.0×10^4 to 4.0×10^4	5.5×10^4 to 6.5×10^4

Example:

What is the elongation of 200 in. of aluminum wire with a 0.01 square in. area if it supports a weight of 100 lb?

Solution:

$$\delta = \frac{P\ell}{AE} \tag{2-8}$$

$$= \frac{(100 \text{ lb}) (200 \text{ in.})}{(0.01 \text{ in.}^2) (1.0 \times 10^7 \text{ lb/in.}^2)}$$

$$\delta = 0.2 \text{ in.}$$

Summary

The important information in this chapter is summarized below.

Young's Modulus Summary

• Hooke's Law states that in the elastic range of a material strain is proportional to stress. It is measured by using the following equation:

$$\delta = \frac{P\ell}{AE}$$

• Young's Modulus (Elastic Modulus) is the ratio of stress to strain, or the gradient of the stress-strain graph. It is measured using the following equation:

$$E = \frac{\sigma}{\varepsilon}$$

STRESS-STRAIN RELATIONSHIP

Most polycrystalline materials have within their elastic range an almost constant relationship between stress and strain. Experiments by an English scientist named Robert Hooke led to the formation of Hooke's Law, which states that in the elastic range of a material strain is proportional to stress. The ratio of stress to strain, or the gradient of the stress-strain graph, is called the Young's Modulus.

EO 1.10	**DEFINE the following terms:**

        a.   **Bulk Modulus**
        b.   **Fracture point**

EO 1.11   **Given stress-strain curves for ductile and brittle material, IDENTIFY the following specific points on a stress-strain curve.**

        a.   **Proportional limit**   c.   **Ultimate strength**
        b.   **Yield point**   d.   **Fracture point**

EO 1.12   **Given a stress-strain curve, IDENTIFY whether the type of material is ductile or brittle.**

EO 1.13   **Given a stress-strain curve, INTERPRET a stress-strain curve for the following:**

        a.   **Application of Hooke's Law**
        b.   **Elastic region**
        c.   **Plastic region**

Elastic Moduli

The *elastic moduli* relevant to polycrystalline material are Young's Modulus of Elasticity, the Shear Modulus of Elasticity, and the Bulk Modulus of Elasticity.

Young's Modulus

Young's Modulus of Elasticity is the elastic modulus for tensile and compressive stress and is usually assessed by tensile tests. Young's Modulus of Elasticity is discussed in detail in the preceding chapter.

Shear Modulus

The *Shear Modulus of Elasticity* is derived from the torsion of a cylindrical test piece. Its symbol is G.

Bulk Modulus

The *Bulk Modulus of Elasticity* is the elastic response to hydrostatic pressure and equilateral tension or the volumetric response to hydrostatic pressure and equilateral tension. It is also the property of a material that determines the elastic response to the application of stress.

Tensile (Load) Tests and Stress–Strain Curves

To determine the load-carrying ability and the amount of deformation before fracture, a sample of material is commonly tested by a *Tensile Test*. This test consists of applying a gradually increasing force of tension at one end of a sample length of the material. The other end is anchored in a rigid support so that the sample is slowly pulled apart. The testing machine is equipped with a device to indicate, and possibly record, the magnitude of the force throughout the test. Simultaneous measurements are made of the increasing length of a selected portion at the middle of the specimen, called the gage length. The measurements of both load and elongation are ordinarily discontinued shortly after plastic deformation begins; however, the maximum load reached is always recorded. *Fracture point* is the point where the material fractures due to plastic deformation. After the specimen has been pulled apart and removed from the machine, the fractured ends are fitted together and measurements are made of the now-extended gage length and of the average diameter of the minimum cross section. The average diameter of the minimum cross section is measured only if the specimen used is cylindrical.

The tabulated results at the end of the test consist of the following.

a.   designation of the material under test

b.   original cross section dimensions of the specimen within the gage length

c.   original gage length

d.   a series of frequent readings identifying the load and the corresponding gage length dimension

e.   final average diameter of the minimum cross section

f.   final gage length

g.   description of the appearance of the fracture surfaces (for example, cup-cone, wolf's ear, diagonal, start)

A graph of the results is made from the tabulated data. Some testing machines are equipped with an autographic attachment that draws the graph during the test. (The operator need not record any load or elongation readings except the maximum for each.) The coordinate axes of the graph are strain for the x-axis or scale of abscissae, and stress for the y-axis or scale of ordinates. The ordinate for each point plotted on the graph is found by dividing each of the tabulated loads by the original cross-sectional area of the sample; the corresponding abscissa of each point is found by dividing the increase in gage length by the original gage length. These two calculations are made as follows.

$$\text{Stress} = \frac{\text{load}}{\text{area of original cross section}} = \frac{P}{A_o} = \text{psi or lb/in.}^2 \tag{2-9}$$

$$\text{Strain} = \frac{\text{instantaneous gage length} - \text{original}}{\text{original gage length}} = \frac{\text{elongation}}{\text{original gage length}} \tag{2-10}$$

$$= \frac{L - L_o}{L_o} = \text{inches per inch x 100} = \text{percent elongation} \tag{2-11}$$

Stress and strain, as computed here, are sometimes called "engineering stress and strain." They are not true stress and strain, which can be computed on the basis of the area and the gage length that exist for each increment of load and deformation. For example, true strain is the natural log of the elongation (ln (L/L_o)), and true stress is P/A, where A is area. The latter values are usually used for scientific investigations, but the engineering values are useful for determining the load-carrying values of a material. Below the elastic limit, engineering stress and true stress are almost identical.

The graphic results, or stress-strain diagram, of a typical tension test for structural steel is shown in Figure 3. The ratio of stress to strain, or the gradient of the stress-strain graph, is called the Modulus of Elasticity or Elastic Modulus. The slope of the portion of the curve where stress is proportional to strain (between Points 1 and 2) is referred to as Young's Modulus and Hooke's Law applies.

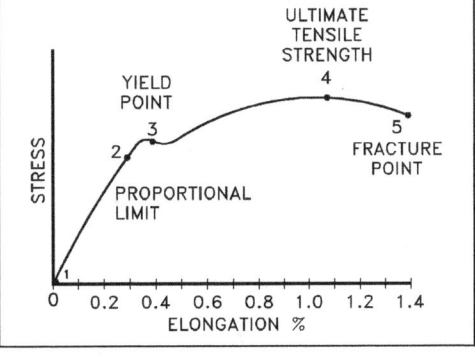

Figure 3 Typical Ductile Material
Stress-Strain Curve

The following observations are illustrated in Figure 3:

- Hooke's Law applies between Points 1 and 2.

- Hooke's Law becomes questionable between Points 2 and 3 and strain increases more rapidly.

- The area between Points 1 and 2 is called the elastic region. If stress is removed, the material will return to its original length.

- Point 2 is the proportional limit (PL) or elastic limit, and Point 3 is the yield strength (YS) or yield point.

- The area between Points 2 and 5 is known as the plastic region because the material will not return to its original length.

- Point 4 is the point of ultimate strength and Point 5 is the fracture point at which failure of the material occurs.

Figure 3 is a stress-strain curve typical of a ductile material where the strength is small, and the plastic region is great. The material will bear more strain (deformation) before fracture.

Figure 4 is a stress-strain curve typical of a brittle material where the plastic region is small and the strength of the material is high.

The tensile test supplies three descriptive facts about a material. These are the stress at which observable plastic deformation or "yielding" begins; the ultimate tensile strength or maximum intensity of load that can be carried in tension; and the percent elongation or strain (the amount the material will stretch) and the accompanying percent reduction of the cross-sectional area caused by stretching. The rupture or fracture point can also be determined.

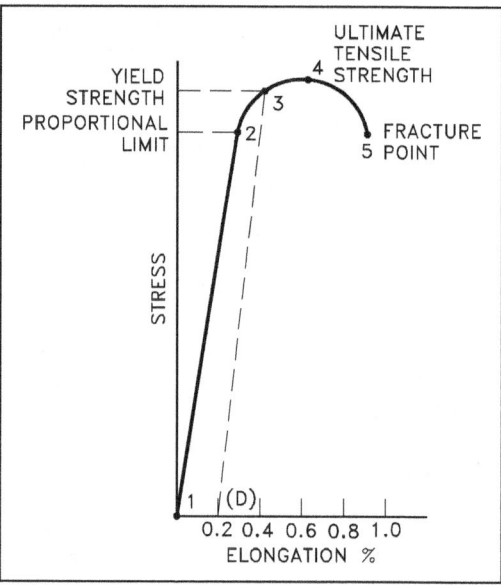

Figure 4 Typical Brittle Material Stress-Strain Curve

Summary

The important information in this chapter is summarized below.

Stress-Strain Relationship Summary

- Bulk Modulus

  The Bulk Modulus of Elasticity is the elastic response to hydrostatic pressure and equilateral tension, or the volumetric response to hydrostatic pressure and equilateral tension. It is also the property of a material that determines the elastic response to the application of stress.

- Fracture point is the point where the material fractures due to plastic deformation.

- Ductile material will deform (elongate) more than brittle material, shown in the figures within the text. The stress-strain curves discussed in this chapter for ductile and brittle demonstrated how each material would react to stress and strain.

- Figures 3 and 4 illustrate the specific points for ductile and brittle material, respectively. Hooke's Law applies between Points 1 and 2. Elastic region is between Points 1 and 2. Plastic region is between Points 2 and 5.

PHYSICAL PROPERTIES

Material is selected for various applications in a reactor facility based on its physical and chemical properties. This chapter discusses the physical properties of material. Appendix A contains a discussion on the compatibility of tritium with various materials.

EO 1.14   DEFINE the following terms:

a.   Strength   d.   Ductility
b.   Ultimate tensile   e.   Malleability
     strength   f.   Toughness
c.   Yield strength   g.   Hardness

EO 1.15   IDENTIFY how slip effects the strength of a metal.

EO 1.16   DESCRIBE the effects on ductility caused by:

a.   Temperature changes
b.   Irradiation
c.   Cold working

EO 1.17   IDENTIFY the reactor plant application for which high ductility is desirable.

Strength

Strength is the ability of a material to resist deformation. The strength of a component is usually considered based on the maximum load that can be borne before failure is apparent. If under simple tension the permanent deformation (plastic strain) that takes place in a component before failure, the load-carrying capacity, at the instant of final rupture, will probably be less than the maximum load supported at a lower strain because the load is being applied over a significantly smaller cross-sectional area. Under simple compression, the load at fracture will be the maximum applicable over a significantly enlarged area compared with the cross-sectional area under no load.

This obscurity can be overcome by utilizing a nominal stress figure for tension and shear. This is found by dividing the relevant maximum load by the original area of cross section of the component. Thus, the strength of a material is the maximum nominal stress it can sustain. The nominal stress is referred to in quoting the "strength" of a material and is always qualified by the type of stress, such as tensile strength, compressive strength, or shear strength.

For most structural materials, the difficulty in finding compressive strength can be overcome by substituting the tensile strength value for compressive strength. This substitution is a safe assumption since the nominal compression strength is always greater than the nominal tensile strength because the effective cross section increases in compression and decreases in tension.

When a force is applied to a metal, layers of atoms within the crystal structure move in relation to adjacent layers of atoms. This process is referred to as *slip*. Grain boundaries tend to prevent slip. The smaller the grain size, the larger the grain boundary area. Decreasing the grain size through cold or hot working of the metal tends to retard slip and thus increases the strength of the metal. Cold and hot working are discussed in the next chapter.

Ultimate Tensile Strength

The *ultimate tensile strength* (UTS) is the maximum resistance to fracture. It is equivalent to the maximum load that can be carried by one square inch of cross-sectional area when the load is applied as simple tension. It is expressed in pounds per square inch.

$$UTS = \frac{\text{maximum load}}{\text{area of original cross section}} = \frac{P_{max}}{A_o} = psi \qquad (2\text{-}12)$$

If the complete engineering stress-strain curve is available, as shown in Figure 3, the ultimate tensile strength appears as the stress coordinate value of the highest point on the curve. Materials that elongate greatly before breaking undergo such a large reduction of cross-sectional area that the material will carry less load in the final stages of the test (this was noted in Figure 3 and Figure 4 by the decrease in stress just prior to rupture). A marked decrease in cross-section is called "necking." Ultimate tensile strength is often shortened to "tensile strength" or even to "the ultimate." "Ultimate strength" is sometimes used but can be misleading and, therefore, is not used in some disciplines.

Yield Strength

A number of terms have been defined for the purpose of identifying the stress at which plastic deformation begins. The value most commonly used for this purpose is the yield strength. The *yield strength* is defined as the stress at which a predetermined amount of permanent deformation occurs. The graphical portion of the early stages of a tension test is used to evaluate yield strength. To find yield strength, the predetermined amount of permanent strain is set along the strain axis of the graph, to the right of the origin (zero). It is indicated in Figure 5 as Point (D).

A straight line is drawn through Point (D) at the same slope as the initial portion of the stress-strain curve. The point of intersection of the new line and the stress-strain curve is projected to the stress axis. The stress value, in pounds per square inch, is the yield strength. It is indicated in Figure 5 as Point 3. This method of plotting is done for the purpose of subtracting the elastic strain from the total strain, leaving the predetermined "permanent offset" as a remainder. When yield strength is reported, the amount of offset used in the determination should be stated. For example, "Yield Strength (at 0.2% offset) = 51,200 psi."

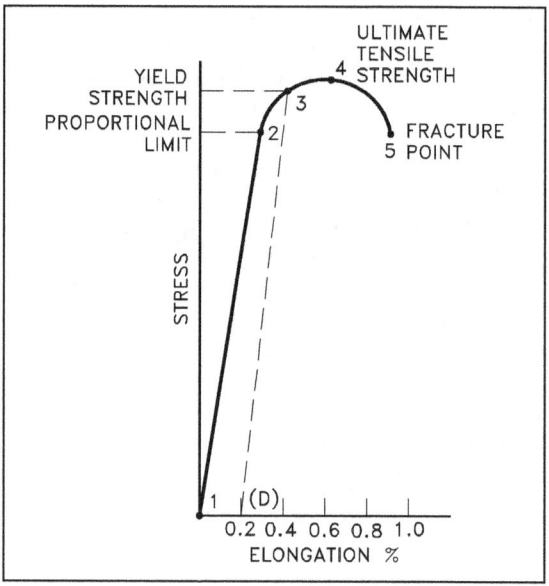

Figure 5 Typical Brittle Material Stress-Strain Curve

Examples of Yield Strength

Some examples of yield strength for metals are as follows.

Aluminum	3.5×10^4 to 4.5×10^4 psi
Stainless steel	4.0×10^4 to 5.0×10^4 psi
Carbon steel	3.0×10^4 to 4.0×10^4 psi

Alternate Values

Alternate values are sometimes used instead of yield strength. Several of these are briefly described below.

- The *yield point*, determined by the divider method, involves an observer with a pair of dividers watching for visible elongation between two gage marks on the specimen. When visible stretch occurs, the load at that instant is recorded, and the stress corresponding to that load is calculated.

- Soft steel, when tested in tension, frequently displays a peculiar characteristic, known as a yield point. If the stress-strain curve is plotted, a drop in the load (or sometimes a constant load) is observed although the strain continues to increase. Eventually, the metal is strengthened by the deformation, and the load increases with further straining. The high point on the S-shaped portion of the curve, where yielding began, is known as the upper yield point, and the minimum point is the lower yield point. This phenomenon is very troublesome in certain deep drawing operations of sheet steel. The steel continues to elongate and to become thinner at local areas where the plastic strain initiates, leaving unsightly depressions called stretcher strains or "worms."

- The *proportional limit* is defined as the stress at which the stress-strain curve first deviates from a straight line. Below this limiting value of stress, the ratio of stress to strain is constant, and the material is said to obey Hooke's Law (stress is proportional to strain). The proportional limit usually is not used in specifications because the deviation begins so gradually that controversies are sure to arise as to the exact stress at which the line begins to curve.

- The *elastic limit* has previously been defined as the stress at which plastic deformation begins. This limit cannot be determined from the stress-strain curve. The method of determining the limit would have to include a succession of slightly increasing loads with intervening complete unloading for the detection of the first plastic deformation or "permanent set." Like the proportional limit, its determination would result in controversy. Elastic limit is used, however, as a descriptive, qualitative term.

In many situations, the yield strength is used to identify the allowable stress to which a material can be subjected. For components that have to withstand high pressures, such as those used in pressurized water reactors (PWRs), this criterion is not adequate. To cover these situations, the maximum shear stress theory of failure has been incorporated into the ASME (The American Society of Mechanical Engineers) Boiler and Pressure Vessel Code, Section III, Rules for Construction of Nuclear Pressure Vessels. The maximum shear stress theory of failure was originally proposed for use in the U.S. Naval Reactor Program for PWRs. It will not be discussed in this text.

Ductility

The percent elongation reported in a tensile test is defined as the maximum elongation of the gage length divided by the original gage length. The measurement is determined as shown in Figure 6.

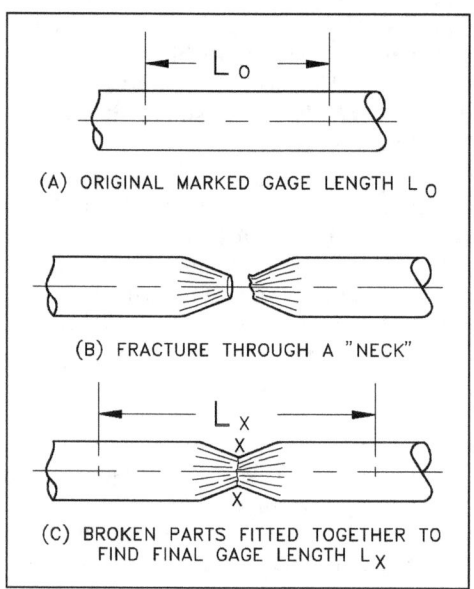

Figure 6 Measuring Elongation After Fracture

$$\text{Percent elongation} = \frac{\text{final gage length} - \text{initial gage length}}{\text{initial gage length}} \qquad (2\text{-}13)$$

$$= \frac{L_x - L_o}{L_o} = \text{inches per inch x 100} \qquad (2\text{-}14)$$

Reduction of area is the proportional reduction of the cross-sectional area of a tensile test piece at the plane of fracture measured after fracture.

Percent reduction of area (RA) =

$$\frac{\text{area of original cross section} - \text{minimum final area}}{\text{area of original cross section}} \qquad (2\text{-}15)$$

$$= \frac{A_o - A_{min}}{A_o} = \frac{\text{decrease in area}}{\text{original area}} = \frac{\text{square inches}}{\text{square inches}} \quad \text{x 100} \qquad (2\text{-}16)$$

The reduction of area is reported as additional information (to the percent elongation) on the deformational characteristics of the material. The two are used as indicators of ductility, the ability of a material to be elongated in tension. Because the elongation is not uniform over the entire gage length and is greatest at the center of the neck, the percent elongation is not an absolute measure of ductility. (Because of this, the gage length must always be stated when the percent elongation is reported.) The reduction of area, being measured at the minimum diameter of the neck, is a better indicator of ductility.

Ductility is more commonly defined as the ability of a material to deform easily upon the application of a tensile force, or as the ability of a material to withstand plastic deformation without rupture. Ductility may also be thought of in terms of bendability and crushability. Ductile materials show large deformation before fracture. The lack of ductility is often termed brittleness. Usually, if two materials have the same strength and hardness, the one that has the higher ductility is more desirable. The ductility of many metals can change if conditions are altered. An increase in temperature will increase ductility. A decrease in temperature will cause a decrease in ductility and a change from ductile to brittle behavior. Irradiation will also decrease ductility, as discussed in Module 5.

Cold-working also tends to make metals less ductile. Cold-working is performed in a temperature region and over a time interval to obtain plastic deformation, but not relieving the strain hardening. Minor additions of impurities to metals, either deliberate or unintentional, can have a marked effect on the change from ductile to brittle behavior. The heating of a cold-worked metal to or above the temperature at which metal atoms return to their equilibrium positions will increase the ductility of that metal. This process is called *annealing*.

Ductility is desirable in the high temperature and high pressure applications in reactor plants because of the added stresses on the metals. High ductility in these applications helps prevent brittle fracture, which is discussed in Module 4.

Malleability

Where ductility is the ability of a material to deform easily upon the application of a tensile force, *malleability* is the ability of a metal to exhibit large deformation or plastic response when being subjected to compressive force. Uniform compressive force causes deformation in the manner shown in Figure 7. The material contracts

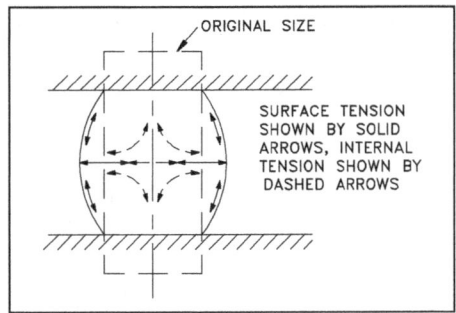

Figure 7 Malleable Deformation of a Cylinder Under Uniform Axial Compression

axially with the force and expands laterally. Restraint due to friction at the contact faces induces axial tension on the outside. Tensile forces operate around the circumference with the lateral expansion or increasing girth. Plastic flow at the center of the material also induces tension.

Therefore, the criterion of fracture (that is, the limit of plastic deformation) for a plastic material is likely to depend on tensile rather than compressive stress. Temperature change may modify both the plastic flow mode and the fracture mode.

Toughness

The quality known as *toughness* describes the way a material reacts under sudden impacts. It is defined as the work required to deform one cubic inch of metal until it fractures. Toughness is measured by the Charpy test or the Izod test.

Both of these tests use a notched sample. The location and shape of the notch are standard. The points of support of the sample, as well as the impact of the hammer, must bear a constant relationship to the location of the notch.

The tests are conducted by mounting the samples as shown in Figure 8 and allowing a pendulum of a known weight to fall from a set height. The

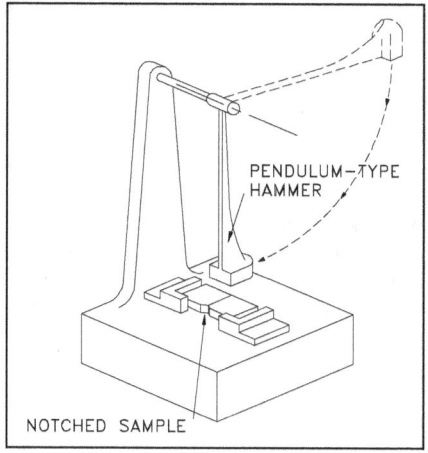

Figure 8 Charpy Test Equipment

maximum energy developed by the hammer is 120 ft-lb in the Izod test and 240 ft-lb in the Charpy test. By properly calibrating the machine, the energy absorbed by the specimen may be measured from the upward swing of the pendulum after it has fractured the material specimen as shown in Figure 9. The greater the amount of energy absorbed by the specimen, the smaller the upward swing of the pendulum will be and the tougher the material is.

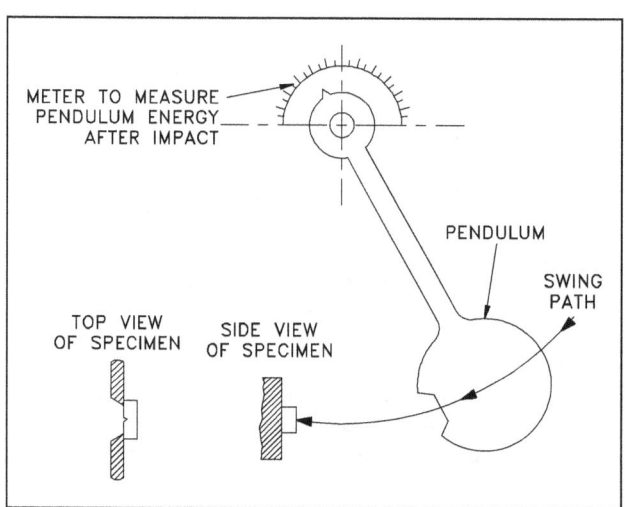

Indication of toughness is relative and applicable only to cases involving exactly this type of sample and method of loading. A sample of a different shape will yield an entirely different result. Notches confine the deformation to a small volume of metal that reduces toughness. In effect, it is the shape of the metal in addition to the material composition that determines the toughness of the material.

Figure 9 Material Toughness Test

Hardness

Hardness is the property of a material that enables it to resist plastic deformation, penetration, indentation, and scratching. Therefore, hardness is important from an engineering standpoint because resistance to wear by either friction or erosion by steam, oil, and water generally increases with hardness.

Hardness tests serve an important need in industry even though they do not measure a unique quality that can be termed hardness. The tests are empirical, based on experiments and observation, rather than fundamental theory. Its chief value is as an inspection device, able to detect certain differences in material when they arise even though these differences may be undefinable. For example, two lots of material that have the same hardness may or may not be alike, but if their hardness is different, the materials certainly are not alike.

Several methods have been developed for hardness testing. Those most often used are Brinell, Rockwell, Vickers, Tukon, Sclerscope, and the files test. The first four are based on indentation tests and the fifth on the rebound height of a diamond-tipped metallic hammer. The file test establishes the characteristics of how well a file takes a bite on the material.

As a result of many tests, comparisons have been prepared using formulas, tables, and graphs that show the relationships between the results of various hardness tests of specific alloys. There is, however, no exact mathematical relation between any two of the methods. For this reason, the result of one type of hardness test converted to readings of another type should carry the notation "_____ converted from _____" (for example "352 Brinell converted from Rockwell C-38").

Another convenient conversion is that of Brinell hardness to ultimate tensile strength. For quenched and tempered steel, the tensile strength (psi) is about 500 times the Brinell hardness number (provided the strength is not over 200,000 psi).

How Alloys Affect Physical Properties

Nickel is an important alloying element. In concentrations of less than 5%, nickel will raise the toughness and ductility of steel without raising the hardness. It will not raise the hardness when added in these small quantities because it does not form carbides, solid compounds with carbon.

Chromium in steel forms a carbide that hardens the metal. The chromium atoms may also occupy locations in the crystal lattice, which will have the effect of increasing hardness without affecting ductility. The addition of nickel intensifies the effects of chromium, producing a steel with increased hardness and ductility.

Copper is quite similar to nickel in its effects on steel. Copper does not form a carbide, but increases hardness by retarding dislocation movement.

Molybdenum forms a complex carbide when added to steel. Because of the structure of the carbide, it hardens steel substantially, but also minimizes grain enlargement. Molybdenum tends to augment the desirable properties of both nickel and chromium.

Stainless steels are alloy steels containing at least 12% chromium. An important characteristic of these steels is their resistance to many corrosive conditions.

Summary

The important information in this chapter is summarized below.

Physical Properties Summary

- Strength is the ability of a material to resist deformation. An increase in slip will decrease the strength of a material.

- Ultimate tensile strength (UTS) is the maximum resistance to fracture.

- Yield strength is the stress at which a predetermined amount of permanent deformation occurs.

- Ductility is the ability of a material to deform easily upon the application of a tensile force, or the ability of a material to withstand plastic deformation without rupture. An increase in temperature will increase ductility. Ductility decreases with lower temperatures, cold working, and irradiation. Ductility is desirable in high temperature and high pressure applications.

- Malleability is the ability of a metal to exhibit large deformation or plastic response when being subjected to compressive force.

- Toughness describes how a material reacts under sudden impacts. It is defined as the work required to deform one cubic inch of metal until it fractures.

- Hardness is the property of a material that enables it to resist plastic deformation, penetration, indentation, and scratching.

WORKING OF METALS

Heat treatment and working of the metal are discussed as metallurgical processes used to change the properties of metals. Personnel need to understand the effects on metals to select the proper material for a reactor facility.

EO 1.18   **STATE how heat treatment affects the properties of heat-treated steel and carbon steel.**

EO 1.19   **DESCRIBE the adverse effects of welding on metal including types of stress and method(s) for minimizing stress.**

Heat Treatment

Heat treatment of large carbon steel components is done to take advantage of crystalline defects and their effects and thus obtain certain desirable properties or conditions.

During manufacture, by varying the rate of cooling (*quenching*) of the metal, grain size and grain patterns are controlled. Grain characteristics are controlled to produce different levels of hardness and tensile strength. Generally, the faster a metal is cooled, the smaller the grain sizes will be. This will make the metal harder. As hardness and tensile strength increase in heat-treated steel, toughness and ductility decrease.

The cooling rate used in quenching depends on the method of cooling and the size of the metal. Uniform cooling is important to prevent distortion. Typically, steel components are quenched in oil or water.

Because of the crystal pattern of type 304 stainless steel in the reactor tank (tritium production facility), heat treatment is unsuitable for increasing the hardness and strength.

Welding can induce internal stresses that will remain in the material after the welding is completed. In stainless steels, such as type 304, the crystal lattice is face-centered cubic (austenite). During high temperature welding, some surrounding metal may be elevated to between 500°F and 1000°F. In this temperature region, the austenite is transformed into a body-centered cubic lattice structure (bainite). When the metal has cooled, regions surrounding the weld contain some original austenite and some newly formed bainite. A problem arises because the "packing factor" (PF = volume of atoms/volume of unit cell) is not the same for FCC crystals as for BCC crystals.

The bainite that has been formed occupies more space than the original austenite lattice. This elongation of the material causes residual compressive and tensile stresses in the material. Welding stresses can be minimized by using heat sink welding, which results in lower metal temperatures, and by annealing.

Annealing is another common heat treating process for carbon steel components. During annealing, the component is heated slowly to an elevated temperature and held there for a long period of time, then cooled. The annealing process is done to obtain the following effects.

      a.   to soften the steel and improve ductility

      b.   to relieve internal stresses caused by previous processes such as heat treatment, welding, or machining

      c.   to refine the grain structure

Cold and Hot Working

Plastic deformation which is carried out in a temperature region and over a time interval such that the strain hardening is not relieved is called *cold work*. Considerable knowledge on the structure of the cold-worked state has been obtained. In the early stages of plastic deformation, slip is essentially on primary glide planes and the dislocations form coplanar arrays. As deformation proceeds, cross slip takes place. The cold-worked structure forms high dislocation density regions that soon develop into networks. The grain size decreases with strain at low deformation but soon reaches a fixed size. Cold working will decrease ductility.

Hot working refers to the process where metals are deformed above their recrystallization temperature and strain hardening does not occur. Hot working is usually performed at elevated temperatures. Lead, however, is hot-worked at room temperature because of its low melting temperature. At the other extreme, molybdenum is cold-worked when deformed even at red heat because of its high recrystallization temperature.

The resistance of metals to plastic deformation generally falls with temperature. For this reason, larger massive sections are always worked hot by forging, rolling, or extrusion. Metals display distinctly viscous characteristics at sufficiently high temperatures, and their resistance to flow increases at high forming rates. This occurs not only because it is a characteristic of viscous substances, but because the rate of recrystallization may not be fast enough.

Summary

The important information in this chapter is summarized below.

Effects of Heat Treatment on Metal Properties Summary

- Quenching

    Varying the rate of cooling (quenching) of the metal controls grain size and grain patterns.

    Grain characteristics are controlled to produce different levels of hardness and tensile strength.

    Hardness and tensile strength increase in heat-treated steel; toughness and ductility decrease.

- Welding

    Produces compressive and tensile stresses

    Stresses are minimized by using heat sink welding and annealing

- Annealing

    Softens steel and improves ductility

    Relieves internal stresses caused by previous processes

    Refines grain structure

CORROSION

Corrosion is a major factor in the selection of material for a reactor plant. The material selected must resist the various types of corrosion discussed in the Chemistry Fundamentals Handbook.

EO 1.20   STATE the reason that galvanic corrosion is a concern in design and material selection.

Corrosion

Corrosion is the deterioration of a material due to interaction with its environment. It is the process in which metallic atoms leave the metal or form compounds in the presence of water and gases. Metal atoms are removed from a structural element until it fails, or oxides build up inside a pipe until it is plugged. All metals and alloys are subject to corrosion. Even the noble metals, such as gold, are subject to corrosive attack in some environments.

The corrosion of metals is a natural process. Most metals are not thermodynamically stable in the metallic form; they want to corrode and revert to the more stable forms that are normally found in ores, such as oxides. Corrosion is of primary concern in nuclear reactor plants. Corrosion occurs continuously throughout the reactor plant, and every metal is subject to it. Even though this corrosion cannot be eliminated, it can be controlled.

General Corrosion

General corrosion involving water and steel generally results from chemical action where the steel surface oxidizes, forming iron oxide (rust). Many of the systems and components in the plant are made from iron.

Some standard methods associated with material selection that protect against general corrosion include:

- The use of corrosion-resistant materials such as stainless steel and nickel, chromium, and molybdenum alloys. (Keep in mind that the corrosion is electrochemical by nature, and the corrosion resistance of the stainless steels results from surface oxide films that interfere with the electrochemical process.)

- The use of protective coatings such as paints and epoxies.

- The application of metallic and nonmetallic coatings or linings to the surface which protects against corrosion, but allows the material to retain its structural strength (for example, a carbon steel pressure vessel with stainless steel cladding as a liner).

Galvanic Corrosion

Galvanic corrosion occurs when two dissimilar metals with different potentials are placed in electrical contact in an electrolyte. It may also take place with one metal with heterogeneities (dissimilarities) (for example, impurity inclusions, grains of different sizes, difference in composition of grains, or differences in mechanical stress). A difference in electrical potential exists between the different metals and serves as the driving force for electrical current flow through the corrodant or electrolyte. This current results in corrosion of one of the metals. The larger the potential difference, the greater the probability of galvanic corrosion. Galvanic corrosion only causes deterioration of one of the metals. The less resistant, more active one becomes the anodic (negative) corrosion site. The stronger, more noble one is cathodic (positive) and protected. If there were no electrical contact, the two metals would be uniformly attacked by the corrosive medium. This would then be called general corrosion.

For any particular medium, a list can be made arranging metals sequentially from most active, or least noble, to passive, or most noble. The galvanic series for sea water is discussed in the Chemistry Fundamentals Handbook.

Galvanic corrosion is of particular concern in design and material selection. Material selection is important because different metals come into contact with each other and may form galvanic cells. Design is important to minimize differing flow conditions and resultant areas of corrosion buildup. Loose corrosion products are important because they can be transported to the reactor core and irradiated.

In some instances, galvanic corrosion can be helpful in the plant. For example, if pieces of zinc are attached to the bottom of a steel water tank, the zinc will become the anode, and it will corrode. The steel in the tank becomes the cathode, and it will not be effected by the corrosion. This technique is known as cathodic protection. The metal to be protected is forced to become a cathode, and it will corrode at a much slower rate than the other metal, which is used as a sacrificial anode.

Localized Corrosion

Localized corrosion is defined as the selective removal of metal by corrosion at small areas or zones on a metal surface in contact with a corrosive environment, usually a liquid. It usually takes place when small local sites are attacked at a much higher rate than the rest of the original surface. Localized corrosion takes place when corrosion works with other destructive processes such as stress, fatigue, erosion, and other forms of chemical attack. Localized corrosion mechanisms can cause more damage than any one of those destructive processes individually. There are many different types of localized corrosion. Pitting, stress corrosion cracking, chloride stress corrosion, caustic stress corrosion, primary side stress corrosion, heat exchanger tube denting, wastage, and intergranular attack corrosion are discussed in detail in the Chemistry Fundamentals Handbook.

Stress-Corrosion Cracking

One of the most serious metallurgical problems and one that is a major concern in the nuclear industry is *stress-corrosion cracking* (SCC). SCC is a type of intergranular attack corrosion that occurs at the grain boundaries under tensile stress. It tends to propagate as stress opens cracks that are subject to corrosion, which are then corroded further, weakening the metal by further cracking. The cracks can follow intergranular or transgranular paths, and there is often a tendency for crack branching.

The cracks form and propagate approximately at right angles to the direction of the tensile stresses at stress levels much lower than those required to fracture the material in the absence of the corrosive environment. As cracking penetrates further into the material, it eventually reduces the supporting cross section of the material to the point of structural failure from overload.

Stresses that cause cracking arise from residual cold work, welding, grinding, thermal treatment, or may be externally applied during service and, to be effective, must be tensile (as opposed to compressive).

SCC occurs in metals exposed to an environment where, if the stress was not present or was at much lower levels, there would be no damage. If the structure, subject to the same stresses, were in a different environment (noncorrosive for that material), there would be no failure. Examples of SCC in the nuclear industry are cracks in stainless steel piping systems and stainless steel valve stems.

The most effective means of preventing SCC in reactor systems are: 1) designing properly; 2) reducing stress; 3) removing critical environmental species such as hydroxides, chlorides, and oxygen; 4) and avoiding stagnant areas and crevices in heat exchangers where chloride and hydroxide might become concentrated. Low alloy steels are less susceptible than high alloy steels, but they are subject to SCC in water containing chloride ions. Nickel-based alloys, however, are not effected by chloride or hydroxide ions.

An example of a nickel-based alloy that is resistant to stress-corrosion cracking is inconel. Inconel is composed of 72% nickel, 14-17% chromium, 6-10% iron, and small amounts of manganese, carbon, and copper.

Chloride Stress Corrosion

One of the most important forms of stress corrosion that concerns the nuclear industry is *chloride stress corrosion*. Chloride stress corrosion is a type of intergranular corrosion and occurs in austenitic stainless steel under tensile stress in the presence of oxygen, chloride ions, and high temperature.

It is thought to start with chromium carbide deposits along grain boundaries that leave the metal open to corrosion. This form of corrosion is controlled by maintaining low chloride ion and oxygen content in the environment and use of low carbon steels.

Caustic Stress Corrosion

Despite the extensive qualification of inconel for specific applications, a number of corrosion problems have arisen with inconel tubing. Improved resistance to caustic stress corrosion cracking can be given to inconel by heat treating it at 620°C to 705°C, depending upon prior solution treating temperature. Other problems that have been observed with inconel include wastage, tube denting, pitting, and intergranular attack.

Summary

The important information in this chapter is summarized below.

Corrosion Summary

- *Corrosion* is the natural deterioration of a metal in which metallic atoms leave the metal or form compounds in the presence of water or gases. General corrosion may be minimized by the use of corrosion-resistant materials and the addition of protective coatings and liners.

- *Galvanic corrosion* occurs when dissimilar metals exist at different electrical potentials in the presence of an electrolyte. Galvanic corrosion may be reduced by the careful design and selection of materials regarding dissimilar metals and the use of sacrificial anodes.

- *Localized corrosion* can be especially damaging in the presence of other destructive forces such as stress, fatigue, and other forms of chemical attack.

- *Stress-corrosion cracking* occurs at grain boundaries under tensile stress. It propagates as stress opens cracks that are subject to corrosion, ultimately weakening the metal until failure. Effective means of reducing SCC are 1) proper design, 2) reducing stress, 3) removing corrosive agents, and 4) avoiding areas of chloride and hydroxide ion concentration.

- *Chloride stress corrosion* occurs in austinitic stainless steels under tensile stress in the presence of oxygen, chloride ions, and high temperature. It is controlled by the removal of oxygen and chloride ions in the environment and the use of low carbon steels.

- Problems occurring with the use of inconel include *caustic stress corrosion cracking*, wastage, tube denting, pitting and intergranular attack. Inconel's resistance to caustic stress corrosion cracking may be improved by heat treating.

HYDROGEN EMBRITTLEMENT

Personnel need to be aware of the conditions for hydrogen embrittlement and its formation process when selecting materials for a reactor plant. This chapter discusses the sources of hydrogen and the characteristics for the formation of hydrogen embrittlement.

EO 1.21   **DESCRIBE hydrogen embrittlement, including the two required conditions and the formation process.**

EO 1.22   **IDENTIFY why zircaloy-4 is less susceptible to hydrogen embrittlement than zircaloy-2.**

Concern

Another form of stress-corrosion cracking is *hydrogen embrittlement*. Although embrittlement of materials takes many forms, hydrogen embrittlement in high strength steels has the most devastating effect because of the catastrophic nature of the fractures when they occur. Hydrogen embrittlement is the process by which steel loses its ductility and strength due to tiny cracks that result from the internal pressure of hydrogen (H_2) or methane gas (CH_4), which forms at the grain boundaries. In zirconium alloys, hydrogen embrittlement is caused by zirconium hydriding. At nuclear reactor facilities, the term "hydrogen embrittlement" generally refers to the embrittlement of zirconium alloys caused by zirconium hydriding.

Sources of Hydrogen

Sources of hydrogen causing embrittlement have been encountered in the making of steel, in processing parts, in welding, in storage or containment of hydrogen gas, and related to hydrogen as a contaminant in the environment that is often a by-product of general corrosion. It is the latter that concerns the nuclear industry. Hydrogen may be produced by corrosion reactions such as rusting, cathodic protection, and electroplating. Hydrogen may also be added to reactor coolant to remove oxygen from reactor coolant systems.

Hygrogen Embrittlement of Stainless Steel

As shown in Figure 10, hydrogen diffuses along the grain boundaries and combines with the carbon (C), which is alloyed with the iron, to form methane gas. The methane gas is not mobile and collects in small voids along the grain boundaries where it builds up enormous pressures that initiate cracks. Hydrogen embrittlement is a primary reason that the reactor coolant is maintained at a neutral or basic pH in plants without aluminum components.

If the metal is under a high tensile stress, brittle failure can occur. At normal room temperatures, the hydrogen atoms are absorbed into the metal lattice and diffused through the grains, tending to gather at inclusions or other lattice defects. If stress induces cracking under these conditions, the path is transgranular. At high temperatures, the absorbed hydrogen tends to gather in the grain boundaries and stress-induced cracking is then intergranular. The cracking of martensitic and precipitation hardened steel alloys is believed to be a form of hydrogen stress corrosion cracking that results from the entry into the metal of a portion of the atomic hydrogen that is produced in the following corrosion reaction.

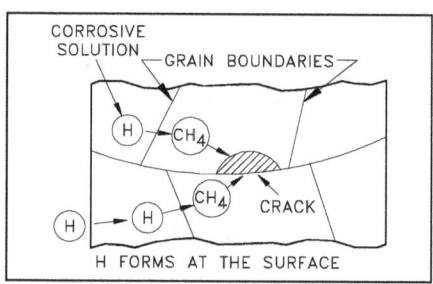

Figure 10 Hydrogen Embrittlement

$$3 \; Fe + 4 \; H_2O \rightarrow Fe_3O_4 + 4 \; H_2$$

Hydrogen embrittlement is not a permanent condition. If cracking does not occur and the environmental conditions are changed so that no hydrogen is generated on the surface of the metal, the hydrogen can rediffuse from the steel, so that ductility is restored.

To address the problem of hydrogen embrittlement, emphasis is placed on controlling the amount of residual hydrogen in steel, controlling the amount of hydrogen pickup in processing, developing alloys with improved resistance to hydrogen embrittlement, developing low or no embrittlement plating or coating processes, and restricting the amount of in-situ (in position) hydrogen introduced during the service life of a part.

Hygrogen Embrittlement of Zirconium Alloys

Hydrogen embrittlement is a problem with zirconium and zirconium alloys, which often are used as cladding materials for nuclear reactors. Zirconium reacts with water as follows.

$$Zr + 2 \; H_2O \rightarrow ZrO_2 + 2H_2\uparrow$$

Part of the hydrogen produced by the corrosion of zirconium in water combines with the zirconium to form a separate phase of zirconium hydride ($ZrH_{1.5}$) platelets. The metal then becomes embrittled (ductility decreases) and it fractures easily. Cracks begin to form in the zirconium hydride platelets and are propagated through the metal. Zircaloy-2 (a zirconium alloy), which has been used as a fuel rod cladding, may absorb as much as 50% of the corrosion-produced hydrogen and is subject to hydrogen embrittlement, especially in the vicinity of the surface. Studies at Westinghouse, Batelle, and elsewhere have revealed that the nickel in the zircaloy-2 was responsible for the hydrogen pickup. This has led to the development of zircaloy-4, which has significantly less nickel than zircaloy-2 and is less susceptible to embrittlement. In addition, the introduction of niobium into zircaloy-4 further reduces the amount of hydrogen absorption.

Summary

The important information in this chapter is summarized below.

Hydrogen Embrittlement Summary

- Hydrogen embrittlement

    The conditions required for hydrogen embrittlement in steel is the presence of hydrogen dissolved in the water and the carbon in the steel. The hydrogen dissolved in the water comes from:

    Making of steel

    Processing parts

    Welding

    Storage or containment of hydrogen gas

    Related to hydrogen as a contaminant in the environment that is often a by-product of general corrosion.

    Hydrogen embrittlement is the result of hydrogen that diffuses along the grain boundaries and combines with the carbon to form methane gas. The methane gas collects in small voids along the grain boundaries where it builds up enormous pressures that initiate cracks and decrease the ductility of the steel. If the metal is under a high tensile stress, brittle fracture can occur.

- Zircaloy-4 is less susceptible to hydrogen embrittlement than zircaloy-2 because:

    Zircaloy-4 contains less nickel

    The introduction of niobium into zircaloy-4 reduces the amount of hydrogen absorption in the metal.

Intentionally Left Blank

Appendix A

Tritium/Material Compatibility

APPENDIX A

TRITIUM/MATERIAL COMPATIBILITY

Concerns

Many compatibility concerns can be raised for tritium/material interactions.

- The mechanical integrity of the material

- The escape rate of tritium into and through the material

- Contamination of tritium by the material and vice versa

- Gettering capabilities of a substance for tritium

Mechanical integrity is a function of how well the material dissipates the energy of colliding beta particles and how well it excludes tritium from its bulk. Cross-contamination occurs when materials contain hydrogen or carbon in their bulk or at their surface or when the materials absorb a significant amount of tritium.

Gettering capabilities are largely a function of alloy overpressure. The process of gettering is the removal of gases by sorption; either adsorption, absorption, or chemisorption. In absorption the atoms of the gas dissolve between the atoms of the alloy. In adsorption and chemisorption, the molecules of the gas adhere to the surface of the alloy. The difference between adsorption and chemisorption is the type and strength of bonds that hold the molecules to the surface.

Compatibility

Because of its radioactive, chemically-reducing, and diffusive properties, tritium gas interacts with almost all materials. Tritium gas permeates and degrades many useful polymeric materials (for example, organics such as pump oils, plastics, and O-rings). This action causes a loss of mechanical properties within months or years.

Tritium gas diffuses through glass, especially at elevated temperatures. The beta rays activate the reduction of Si-O-Si bonds to Si-OT and Si-T bonds, and mechanical properties may be lost over a period of years.

Some metals, such as uranium, are directly hydrided by tritium gas. These metals form a chemical compound and their mechanical properties are altered within minutes or hours. However, some metals, such as stainless steels, are permeated by tritium, but do not lose their mechanical properties unless the tritium pressure is hundreds of atmospheres for several years.

Solubility in Metals

Hydrogen dissolves as atoms in metals. These atoms occupy octahedral and tetrahedral locations within the lattice. The hydrogen apparently exists within nonhydriding metal lattices as proton, deuteron, or triton, with the electron in a metal conduction band. Some metals are endothermic (chemical change due to absorption of heat) hydrogen absorbers and others are exothermic (chemical change that releases heat), and solubilities vary considerably (approximately 10 to 15 orders of magnitude) at room temperature.

The solubility of hydrogen in endothermic absorbers increases as the temperature increases. The reverse is true for exothermic absorbers and the solubility decreases as the temperature increases. For various hydride phases, plots of decomposition overpressure as a function of inverse temperature yield negative enthalpies or heats of formation.

Permeability

Permeability (Φ) of gas (including H_2 or T_2) through materials is a measure of how much gas will migrate across a material wall of given thickness and area over a given time. It is a direct function of the ability to diffuse and solubility. Dimensionally,

$$\Phi \left(\frac{cm^3(H_2,\ STP)\ \cdot\ cm(thickness)}{cm^2(area)\ \cdot\ sec.} \right) = D \left(\frac{cm^2}{sec.} \right) \cdot S \left(\frac{cm^3(H_2,\ STP)}{cm^3(material)} \right) \qquad (A-1)$$

where:

Φ = permeability

D = diffusivity

S = solubility

The following materials are listed in order of increasing permeability: ceramics and graphite, silicas, nonhydriding metals, hydriding metals, and polymers. The permeability of many other hydrogen-bearing molecules through polymers has been studied. For such molecules, permeability can be well in excess of that for hydrogen through a polymer. This must be considered when handling tritiated water or organic solvents.

Two factors that influence the permeability of a metal are oxides on surface and surface area. Because the permeability of hydrogen through a metal oxide at a given temperature is usually orders of magnitude lower than it is through the metal, a thin surface oxide can markedly reduce the permeability of hydrogen through the material.

For example, if LiD salt is placed in contact with the surface of a stainless steel specimen, the oxide is reduced, allowing increased permeation. If a metal undergoes surface oxidation in the presence of steam, permeability decreases as oxidation proceeds.

Nonhydriding Metals

The mechanical integrity of nonhydriding metals in the presence of tritium is excellent because the electron bands carry away the energy of colliding beta particles without disrupting the metal structure or bonding. These metals form the most common class of tritium containment structural materials. They generally include 304L, 316L, 321, 21-6-9, and Nitronic stainless steels, as well as copper and aluminum. Inconel, Ni-Cr alloys, and 400-series stainless steels are generally not chosen because of corrosion or hydrogen embrittlement sensitivity. At high pressures of tritium gas, however, classical hydrogen embrittlement, as well as helium-3 embrittlement, can occur in accepted materials. For example, for 304L stainless steel samples exposed to 9 kpsi of tritium at 423 K for 6 months and then aged 1.5 years, fracture toughness decreased by a factor of 6. Of this, a factor of two could be attributed to helium-3 alone.

Substantially different fracture modes are observed between aged tritium-loaded and unloaded steel specimens. Helium-3 is vastly less soluble in metals than is hydrogen (tritium); helium pockets (bubbles) form with high internal pressures. Hydrogen embrittlement also contributes to this effect.

Permeative escape rates of tritium through nonhydriding metals are generally acceptable at temperatures below 100°C to 300°C and for thicknesses of 0.1 cm or more. For 304 stainless steel 0.3 cm thick with a 1000-cm^2 surface area exposed on one side to tritium gas of 1 atm pressure at 300 K, the permeability is 1.6 x 10^{-4} Ci/day ($t_{0.9}$ = 7 hours). The temperature dependence of permeation is often astounding.

Cross-contamination between nonhydriding metals and tritium does occur often enough to be troublesome. Oxide layers on metals often contain hydrogen and are further covered with a thin adsorbed carbonaceous film when originally grown in room air. Upon exposure to such a surface, tritium gas may become contaminated over hours or days with hundreds to thousands of parts per million of protium (as HT) and methane (as CT_4) as the surface layers are radiolyzed, exchanged, and contaminated by the material. Because diffusion of tritium in the bulk material is usually slow at room temperature, the extent of surface oxide contamination may greatly surpass the bulk contamination of a component. Cross-contamination can be minimized by minimizing material surface areas, choosing an impermeable material with a thin or nonexistent oxide layer, and maintaining cleanliness.

Tritium present in an oxide layer can be removed by acid dissolution of the oxide or more gently by isotopic exchange with normal water or activated hydrogen gas (plasma). Because diffusion of oxide- or bulk-dissolved tritium back to the surface of a material undergoing decontamination is often slow, exchange at an elevated temperature may be advantageous.

Hydriding Metals

When exposed to tritium gas, hydriding metals absorb large volumes of tritium to form tritide phases, which are new chemical compounds, such as UT_3. The mechanical integrity of the original metallic mass is often severely degraded as the inclusions of a brittle, salt-like hydride form within the mass. Because of this property and their large permeability to hydrogen, hydriding metals are not to be used for constructing pipelines and vessels of containment for tritium gas. They have great utility, however, in the controlled solidification and storage of tritium gas, as well as in its pumping, transfer, and compression.

Uranium, palladium, and alloys of zirconium, lanthanum, vanadium, and titanium are presently used or are proposed for pumping and controlled delivery of tritium gas. Several of these alloys are in use in the commercial sector for hydrogen pumping, storage, and release applications. Gaseous overpressure above a hydride (tritide) phase varies markedly with temperature; control of temperature is thus the only requirement for swings between pumping and compressing the gas.

In practice, pumping speeds or gaseous delivery rates (the kinetic approach to equilibrium) are functions of temperature (diffusion within the material), hydride particle size, and surface areas and conditions. Poisoning of a uranium or zirconium surface occurs when oxygen or nitrogen is admitted and chemically combines to form surface barriers to hydrogen permeation. In practice, these impurities may be diffused into the metal bulk at elevated temperature, thereby reopening active sites and recovering much of the lost kinetics. Other metals and alloys (for example, $LaNi_3$) are less subject to poisoning, although alloy decomposition can occur.

Helium-3, generated as microscopic bubbles within the lattice of tritides, is not released except by fracture and deformation of metal grains. This release usually occurs at high temperature or after long periods of time. When a tritide is heated to release tritium, helium-3 is also released to some extent. The cooled metal, however, does not resorb the helium-3. The practice of regenerating a tritide storage bed to remove helium-3 immediately prior to use for pure tritium delivery is therefore common.

If helium-3 (or another inert impurity) accompanies tritium gas that is absorbed onto a tritide former, helium blanketing may occur. The absorption rate slows as the concentration of helium in the metal crevices leading toward active sites becomes high. Normal gaseous diffusion is often not sufficient to overcome this effect. Forced diffusion by recirculating the gas supply can be used to overcome blanketing.

Graphite

Because they generally have high surface areas, graphite samples adsorb large amounts of hydrogen gas (4×10^{18} molecules/g for a graphite pellet used in gas-cooled reactors). Methane, protium, and (possibly) water are generated from beta irradiation of the graphite surface.

The surface of the graphite will be contaminated with chemically-bound tritium, and decontamination may be possible by baking the graphite at 500°C in the presence of a hydrogen exchange medium, such as H_2, H_2O, or NH_3. Except for possible surface erosion, graphite will probably not be degraded mechanically even over a period of several years, as bulk diffusion and solubility are extremely low.

Glasses

Various data suggest that tritium gas in the presence of its chemically-activating beta irradiation energy could reduce silica bonding to -Si-T and -Si-OT species. At temperatures above 300°C, deuterium appears to reduce silica network, and dissolved deuterium in a gamma irradiation field has the same effect. The migration of tritium into glass structures could, therefore, cause embrittlement and possibly fracture under stress over several months or years. Evidence also suggests that activated hydration of glassy silica structures under T_2O exposure is possible. Embrittlement (unexpected fracture) of a Pyrex syringe stored for two to three years after being used to transfer T_2O was experienced at one DOE nuclear facility.

Permeability of silica glasses is one to two orders of magnitude greater than that for stainless steel over the temperature range 0° to 200°C. Tritium-handling systems constructed largely of glass have nevertheless been widely used, although this material is not in favor today except for tritium lamp containment. The exchange of tritium with naturally occurring hydroxyl groups in various glasses and on their surfaces is a source of protium contamination to tritium, perhaps 1% HT into 1 atm tritium within a 1-L glass container after 1 year. Decontaminating a highly-exposed glass of its bound tritium would require a significant number of water washes of 300°C hydrogen permeation flushes. This effort is likely to be costly and is often not warranted by the value of the part undergoing decontamination.

Ceramics

Because tritium's solubility, ability to diffuse, and permeability are so much lower for ceramics than for glasses, ceramics undergo little or no bulk disruption from tritium. However, some mechanical degradation of regions near the surface is possible. The depth of the area affected is a function of ability to diffuse and time. Oxygen release from Al_2O_3 (sapphire) windows in the presence of liquid T_2O has recently been noted, although compatibility with tritium gas has been described as excellent. The exchange of surface and near-surface protium is likely, although mutual contamination of tritium and the ceramic should be less than that for glasses. Tritium-contaminated ceramics can probably be decontaminated by warm water or steam flushes or by etching in an acidic solution.

Plastics, Elastomers, and Oils

Organics are easily permeated by tritium (gas or water) and are therefore subject to disruption of their bulk chemistries. There are few or no mechanisms for rapidly delocalizing beta energy, and substantial mobility of organic chains occur within polymer structures (particularly amorphous regions). Once formed, reactive organic intermediates can thus react with each other.

These effects are important when considering the design of tritium systems. Damage to components, such as gaskets, valve tips, and O-rings, must be carefully considered. Component failure during service can cause a major release of tritium. Because elastomer seals often become embrittled, maintenance on nearby sections of piping may cause seals to develop leaks as the result of mechanical movement in the seal area.

Figure A-1 illustrates several polymer chain modifications that take place following activation by beta radiation to ionic or excited species. Cross-linking and degradation are the most important processes to the mechanical properties of the polymer. These both compete in a material, but those polymers that are most sterically hindered appear to preferentially degrade. Steric hindrance prevents neighboring chains from linking and also imparts structural strains that are relieved upon chain scissioning. Cross-linking is noted mechanically by an increase in tensile strength and a decrease in elongation, whereas degradation is evidenced by a decrease in tensile strength, an increase in elongation, and softening of the polymer to a gummy consistency.

Several factors effect polymer stability. First, energy-delocalizing aromatic structural groups increase polymer stability by distributing energies of excited states. In addition, halogen atoms within polymers generate free radicals and thus promote radiation damage.

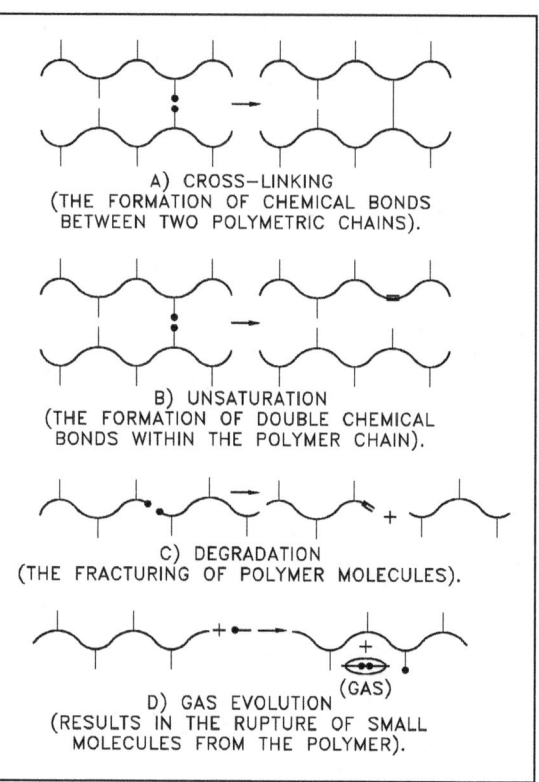

A) CROSS-LINKING
(THE FORMATION OF CHEMICAL BONDS BETWEEN TWO POLYMETRIC CHAINS).

B) UNSATURATION
(THE FORMATION OF DOUBLE CHEMICAL BONDS WITHIN THE POLYMER CHAIN).

C) DEGRADATION
(THE FRACTURING OF POLYMER MOLECULES).

D) GAS EVOLUTION
(RESULTS IN THE RUPTURE OF SMALL MOLECULES FROM THE POLYMER).

Figure A-1 Modifications to Polymer Chains
Due to Irradiation

Substituents on aromatic groups that extend the delocalized bonding network are further stabilizers. Finally, saturated aliphatics are more radiation resistant than those that are unsaturated; isolated double bonds are readily excited to ions or radicals.

Organic compounds, in order of decreasing radiation resistance, are aromatics, aliphatics, alcohols, amines, esters, ketones, and acids. Extension to beta radiation is probably reasonable. In tritium gas, however, substantial differences in irradiation or polymer surface as compared to bulk can occur. This results from the greater density of tritium (and the much greater range of the beta in the tritium gas) outside the polymer compared to inside the polymer bulk.

Some direct experience of polymers with tritium has been obtained. Teflon, Viton, or Kel-F exposure in tritium produces the acid TF, noted as SiF_4, gas in a glass system. Because of this acid production, tritium + moisture + Teflon in a stainless steel system at pressures of approximately 1300 atm caused catastrophic stress corrosion cracking of 0.76-mm thick stainless steel tube walls in 16 hours. Substituting deuterium for tritium or removing Teflon or moisture caused no failure. Radiation damage to Teflon is more severe than to all other thermoplastics. Teflon is therefore not recommended in the presence of concentrated tritium streams.

Surface and bulk effects have been noted in numerous polymer/tritium studies. In one study, hardening of neoprene occurred throughout the bulk, while hardening of natural rubber primarily occurred at the surface (crack propagation). Total incorporation of tritium into a polyethylene powder was found not to be a function of the amount of powder, but of the exposed surface area. Radiation-induced fluorescence from the surface of high-density polyethylene exposed to tritium was shown to be orders of magnitude greater than that from the bulk.

Polyimides (good in the presence of gamma radiation) appear good in tritium handling and are recommended. Vespel stem tips for valves, when used with sufficient sealing force, continue to seal for several years in tritium (STP). When used with less sealing force, however, leaks have been noted across valve tips, possibly because of surface hardening. Polyimide gaskets under constant sealing load are probably adequate for years.

Saturated hydrocarbon mineral oils (for example, Duo-Seal) require frequent changes in tritium service because of vapor pressure increases (offgassing) and liquid viscosity increases. Silicone oils are rapidly polymerized or solidified. Polyphenyl ether oils last for years in similar service, but are expensive and may absorb significant amounts of tritium.

Fluorinated pump oils are not recommended for tritium service and certainly not for tritiated water vapor service. Tritium fluoride evolution and corrosion may result.

www.ingramcontent.com/pod-product-compliance
Lightning Source LLC
Chambersburg PA
CBHW082139290526
45794CB00008B/3098